Structure of the Schrödinger Dynamics

Kazuo Takatsuka

Structure of the Schrödinger Dynamics

Rediscovery of Quantum Mechanics

Springer

Kazuo Takatsuka
Fukui Inst. for Fundamental Chemistry
Kyoto University
Kyoto Sakyo, Japan

ISBN 978-981-92-1909-4 ISBN 978-981-92-1910-0 (eBook)
https://doi.org/10.1007/978-981-92-1910-0

The original submitted manuscript has been translated into English. The translation was done using artificial intelligence. A subsequent revision was performed by the author(s) to further refine the work and to ensure that the translation is appropriate concerning content and scientific correctness. It may, however, read stylistically different from a conventional translation.

This Springer imprint is published by the registered company Springer Nature Singapore Pte Ltd.
The registered company address is: 152 Beach Road, #21-01/04 Gateway East, Singapore 189721, Singapore

Preface

The purpose of this book is to further develop and extend novel structures of the fundamental framework of quantum theory beyond those already established as presented in the classic works of Dirac, Messiah, Landau–Lifshitz, and so on. In particular, this work aims to complement and extend Schrödinger dynamics, focusing on developments in quantum mechanics that are not addressed in canonical textbooks such as those by Dirac.

The year 2025 marks the centennial of the birth of quantum mechanics. Under the banner of "quantum supremacy," a new era is unfolding, encompassing quantum technologies, quantum information, and quantum metrology. Quantum mechanics is a peculiar theory: it is regarded as a "complete theory" while simultaneously being "imperfect." Its level of completion is remarkably high; centered on the Schrödinger equation and the wavefunction, it forms a grand and coherent theoretical edifice incorporating functional analysis, operator algebras, and so on. Furthermore, approximation methods such as perturbation theory and variational principles are highly developed, and within its wide range of applications, no empirical facts contradicting quantum theory have been found (within its domain of validity, such as the non-relativistic regime).

On the other hand, it is fair to say that there remains considerable confusion regarding the fundamental understanding and interpretation of quantum theory. This is precisely why quantum theory is described as "imperfect." Representative issues include the following: (1) What is the wavefunction? What does the Schrödinger equation represent, how is it derived, and what is the physical meaning and role of the imaginary unit appearing in its fundamental formulation? (2) Regarding Born's probability interpretation, Einstein famously objected that "God does not play dice," rejecting the introduction of probability into fundamental principles. Although the probability interpretation has been extraordinarily successful in practice, what is its physical origin? (3) Does the wavefunction represent a "matter wave"? Does it collapse at the moment of observation? Conversely, are interpretations proposed to avoid collapse valid? (4) In the double-slit experiment, through which path does an electron travel to produce the interference pattern? (5) Einstein cast a doubt on the uncertainty relation based on his claimed concept "physical reality." Then he and his colleague argued that long-range entanglement can indeed be exploited to violate the uncertainty relation. Moreover, since the entanglement and uncertainty relation seem to be mutually incompatible within quantum mechanics, they came to conclusion that quantum theory is either incorrect or incomplete. The 2022 Nobel Prize in Physics awarded to John Clauser, Alain Aspect, and Anton Zeilinger for experimentally demonstrating violations of Bell's inequality was even interpreted as settling a century of confusion regarding the quantum entanglement. However, does the violation of Bell's inequality truly constitute a refutation to the doubt cast by Einstein? Have we genuinely resolved the objections, doubts, and skepticism he posed and moved beyond them? Einstein's concerns addressed the very foundations upon which scientific laws, including quantum theory, are built. In that sense, they are fundamental—indeed, seemingly elementary—questions. Yet these are precisely the issues that remain unclear even after studying canonical texts; according to Bohr (arguably), one is instead urged to refrain from such contemplation and focus solely on calculation.

In fact, there is a reason for this "incompleteness." It arises from attempting to view the entirety of quantum theory within the axiomatic principle, monolithic structure of the (non-relativistic) Schrödinger equation. If one asks questions that the Schrödinger equation is capable of answering, it provides correct answers. Classic texts such as Dirac's exemplify this systematic approach in a beautiful fashion. Yet, quantum mechanics contains additional theoretical structures beyond this framework. These may be summarized in the following two points: (1) the logical structure underlying the origin of the Schrödinger equation. This has been recognized as a profound issue since the birth of quantum mechanics. Like any theory, the Schrödinger equation does not contain within itself a self-referential account of its own derivation. But still, it is insufficient to rely on analogies, like the Hamilton-Jacobi theory; a logically compelling construction is required. Such an approach should clarify the physical meaning of the wavefunction, the role of the Schrödinger equation, the dynamical significance of complex numbers, and the origin of the superposition principle. (2) The Schrödinger wavefunction is, in essence, a distribution amplitude describing an ensemble of events. The individual events comprising this ensemble—corresponding to each single measurement outcome—should possess their own equations of motion and theoretical background. Confusing properties of individual events with those of the entire wavefunction may lead to seemingly paradoxical conclusions.

This book provides a mathematical discussion of these underlying structures supporting quantum mechanics. More specifically, the "structure of Schrödinger dynamics" considered here refers to the real-valued Schrödinger equation derived from first principles, a real-valued matrix-form path integral with an integral measure serving as its Green's function, and quantum stochastic differential equations describing the dynamics of individual events along quantum paths. Interpretations of quantum theory are also treated within this framework.

The quantum theory presented in this book is not intended as a theory for theory. In the century following its birth, emerging quantum technologies will increasingly advance toward experimental investigations that observe the dynamics of individual quantum systems. In such contexts, it will be necessary not only to avoid the confusion arising from an exclusive reliance on the Schrödinger equation but also to acquire and expand new perspectives. If the extended framework developed here can deepen our understanding of quantum theory and provide a theoretical foundation for new experimental research in quantum science, it will be a source of great delight.

Kyoto, Japan Kazuo Takatsuka

Acknowledgments The three-pillar structure of Schrödinger dynamics developed in this book from Chap. 5 onward is entirely based on the author's original works. The quantum stochastic differential equation presented in Chap. 8 is also due to the author; however, in this regard, the author has been greatly influenced by the pioneering works of Masao Nagasawa (Professor Emeritus, University of Zurich) on the "theory of quantum random motion." The author would like to express sincere respect and gratitude here.

The author has long been engaged in applied research in quantum mechanics, particularly concerning the Schrödinger equation, including quantum scattering theory, quantum chaos, nonadiabatic electron dynamics, and semiclassical mechanics. Among these, semiclassical theory, in addition to its role as an approximation theory, is a field that strongly compels one to consider the question, "What makes quantum mechanics quantum?" This book represents the author's own attempt to summarize a certain conclusion regarding the nature of quantum theory. The author would like to express sincere thanks to Dr. Satoshi Takahashi, with whom many years of research in semiclassical theory have been shared.

This book was originally drafted in Japanese and translated into English by ChatGPT under the license provided to Springer Nature. The first English edition has been corrected, augmented and polished by the present author. It is this author who is responsible for the final English edition.

Competing Interests The author has no competing interests to declare that are relevant to the content of this manuscript.

Contents

1 Introduction

Abstract

After 100 years of quantum mechanics, this book introduces and guides to quantum theory that are virtually not known by most of scientists. And, this book is written for those who routinely work with quantum theory—particularly the Schrödinger equation—as well as for students and researchers specializing in quantum information and quantum technology, who wish to reconsider the fundamental question: What is Schrödinger dynamics? To address this fundamental question, we propose three-pillared structure of the dynamics:

1. The real-valued Schrödinger equation and wavefunction derived from scratch,
2. The real-valued path integral as its Green function,
3. Stochastic differential equation for quantum path dynamics.

Concrete and physical (not philosophical) interpretations follow the present mathematical formalism, such as

- The physical meaning and role of the Schrödinger equation,
- Comprehension of the particle–wave duality and Born's probabilistic interpretation,
- Denial of wavepacket collapse,
- Mechanism of formation of the fringe (interference) pattern in the double-slit experiment,
- Resolving EPR paradox by finding the spontaneous disentanglement,
- Kinematic origin of the uncertainty relation,
- Quantum-mechanical Hamilton canonical equations of motion, and more.

K. Takatsuka, *Structure of the Schrödinger Dynamics*,
https://doi.org/10.1007/978-981-92-1910-0_1

Keywords

Real-valued Schrödinger equation · Schrödinger vector · Real-valued path integral · Quantum stochastic differential equation · Entanglement · Wavepacket collapse · Double-slit experiment

The Schrödinger equation was born abruptly in the mind of the genius Erwin Schrödinger while he was investigating the origin of the discrete spectrum of the hydrogen atom under the image of de Broglie's matter waves. It was a kind of "inspiration" that descends only upon those who are able to think deeply and persistently. Yet even Schrödinger himself did not fully understand the true meaning of the equation, and it was mysterious enough that he, at times, expressed views that were later regarded as incorrect.

Among the group of geniuses who emerged during this revolutionary period, Heisenberg, Schrödinger's rival, was particularly philosophical in temperament. Einstein held a firm conviction that "scientific laws and theories must be of a certain form," while Bohr, who confronted Einstein, was a profound thinker and contemplative philosopher. As a result, discussions concerning the essence of quantum theory grew ever deeper and at the same time more chaotic. Even today, rather than converging toward a consensus on its understanding and interpretation, the situation is closer to a state of a hundred schools of thought contending.

On the other hand, it is well known that the guiding principle often attributed to Bohr—"Do not think; just calculate"—has been remarkably successful for nearly one hundred years. Indeed, in the classroom, we ourselves learn that classical mechanics, as recognized from the late nineteenth to the early twentieth century, breaks down (or more precisely, reaches the limit of its applicability) and that a new mechanical framework is required, culminating in de Broglie's concept of matter waves. We then experience how the Schrödinger equation appears as the theory that resolves these difficulties. After learning the basic formalism of the Schrödinger equation and how to understand the solutions of typical equations, we gradually enter a vast world of applications. This is the usual structure of a standard course in quantum mechanics.

Through repeated successes in these applications, we gain confidence in the correctness of quantum theory. At the same time, we tend to forget the naive questions we once had in our youth: What exactly is the Schrödinger equation? Where does it come from? How can a fundamental law of nature involve imaginary numbers?

In research and education settings where quantum mechanics is applied, the de facto standard understanding of "what quantum theory is" may be summarized as follows: "In quantum mechanics, the state of motion of light particles such as electrons is completely described by a wavefunction. The wavefunction obeys the Schrödinger equation. The Schrödinger equation is a linear partial differential equation that admits wave-like solutions, making it possible to describe the wave nature of particles as matter waves. It satisfies the principle of superposition and, depending on boundary conditions, yields discrete energy eigenvalues, thereby describing phenomena characteristic of quantum systems." To this is usually added: "The square of the absolute value of the wavefunction represents the probability that a phenomenon will occur." These statements may be regarded as a general consensus concerning quantum theory. Yet, in fact, they provide no physical foundation whatsoever. Nevertheless, this doctrine has been steadily accepted and applied for a century in research environments where quantum theory is used and in educational environments that prepare people to enter such fields. Among the terms appearing above, the only one whose meaning may be relatively clear even to beginners is probably "light particles such as electrons." The circular definition between the wavefunction and the Schrödinger equation, which is reworded even in a more simplistic form as "The wavefunction obeys the Schrödinger equation and the Schrödinger equation drives the wavefunction," may read to advising us to sufficiently and successfully apply to scientific problems, and then, we will eventually cease to be bothered by it. However, in order to reveal the essential structure of quantum mechanics, it is precisely this circular definition between the Schrödinger equation and its wavefunction that must be resolved.

It is said that Schrödinger himself spent his entire life seeking for the true meaning of his found equation. In this book, we attempt to follow that path to the origin, albeit by a different route. Accordingly, this book is not an elementary introduction or textbook on quantum theory. Rather, it corresponds to an academic monograph composed of original content. There already exist many celebrated textbooks on quantum mechanics. The present book is concerned with the major questions that remain after studying those major textbooks.

That said, this book may still be read as a (somewhat advanced) textbook, starting from Chap. 4: "Stochastic Processes: Diffusion Equation, Feynman-Kac Formula, and Ito Stochastic Differential Equation". For readers who wish to start from the very beginning, Chap. 2: "Purpose and Outline of This Book" briefly explains the objectives and outline of the book. Next, in Chap. 3: " Schrödinger Equation: The Location of the Problem," we review quantum mechanics at the undergraduate level and, along with a basic explanation of quantum mechanics, provide a brief introduction to what the main issues are. I believe that even full users of quantum mechanics, but not thoroughly familiar with the fundamentals of it, will understand the motivation of this book. Quantum mechanics and diffusion theory are, in fact, closely related in terms of mathematical structure, and their relationship is closer than it might appear at first glance. Therefore, Chap. 4: "Stochastic Processes: Diffusion Equation, Feynman-Kac Formula, and Ito Stochastic Differential Equation" covers the necessary level of diffusion theory (stochastic process theory) for this book. The three-pillared structure of quantum mechanics can be seen as quite similar to the three-pillar structure of diffusion theory. Those who studied before diffusion theory in detail may want to skip this chapter. Continuing, the main topics are Chap. 5 and the following four chapters.

Chapter 5: "Real-Valued Schrödinger Equation: Rebuilding from Scratch." Here, we derive the Schrödinger equation from scratch and clarify its physical meaning.

In doing so, we break the circular definitions between the Schrödinger equation and Schrödinger wavefunction.

Chapter 6: "Real-Valued Path Integral as Green's Functions of the Schrödinger Equation." We express the path integral for the Green's function of the Schrödinger equation in terms of real variables, introducing the path integral measure. (Since Feynman's original path integral is written in terms of complex elements, the integral measure required is not strictly defined.)

Chapter 7: "Summation over Events and Distribution Functions." We reconsider what distribution functions mean in quantum theory, including the problem of "Schrödinger cat."

Chapter 8: "Quantum Stochastic Paths and Stochastic Differential Equations." We write down the quantum stochastic path dynamics equation that describes individual quantum events as a constituent of the ensemble. What "individual quantum events" means is explained in Chap. 2: "Purpose and Outline of This Book."

Chapter 9: "The Nature and Role of Stochastic Paths in Quantum Mechanics." The meaning of the concept of trajectory in quantum theory is to clarify various quantum-theoretical physical effects and phenomena that arise due to the inclusion of the Wiener process, which is a stochastic process, in quantum probability theory and quantum dynamics, some of the examples being the possible origin of the uncertainty relations, the spontaneous disentanglement, and so on.

Finally, Chap. 10 concludes this book summarizing on what are achieved and what remain unresolved as the ultimate mystery of quantum mechanics.

It has been one hundred years since quantum mechanics was established. While our understanding of quantum theory remains in a state of conceptual confusion, we have now entered an era focused on how to further develop quantum technologies and achieve quantum supremacy. However, not only in quantum theory, but also in any scientific discipline, it is precarious to merely manipulate given algorithms without attaining a deep understanding of their foundations. And this book is written for those who routinely work with quantum theory—particularly the Schrödinger equation—as well as for students and researchers specializing in quantum information and quantum technology, who wish to reconsider the fundamental question: What is Schrödinger dynamics? It is intended for readers who aspire to reach a renewed and deeper understanding. The set of mathematical formulations proposed herein is expected to serve as a foundation for research practice and, furthermore, to be extended and developed in future work.

On the other hand, as is well known, quantum mechanics is a field in which philosophical interpretations and metascientific discussions have flourished, precisely because of the difficulty of its conceptual understanding. The many-worlds interpretation, Schrödinger's cat, the instantaneous collapse of the wave packet, and the peculiar phenomenon of entanglement—seemingly capable of transmitting correlations instantaneously across the universe—are among the most frequently cited examples. Some of these ideas appear extravagant, and many seem unlikely to contribute directly to practical scientific work. Nevertheless, as exemplified by the questions posed by Einstein, there remain numerous issues that demand genuine confrontation. In this sense, quantum theory is still both stimulating and profoundly challenging.

In contrast, the present book seeks to elucidate the structure of Schrödinger dynamics in a mathematically precise manner, avoiding purely abstract speculation. More concretely, it attempts to extend a theoretical framework that is eventually usable in practice. As aforementioned, the main subject of this book begins with Chap. 5. The preceding chapters are written in a textbook-like or expository style in order to prepare the necessary awareness of the problems—namely, to clarify what the problems are and what this work aims to resolve.

The author is not a historian of science and is fully aware that the discussions presented up to Chap. 3 regarding classical interpretations of quantum theory and their underlying philosophical backgrounds may be subject to possible criticism as being insufficiently rigorous. In particular, I confess that many of the well-known "statements" and "anecdotal accounts" referred to in this book have not been traced back to original papers or primary sources. On the other hand, from Chap. 5 onward, the interpretations advocated in this book are always accompanied by explicit mathematical formulations, and in that sense, they should be regarded as a form of original "scientific interpretation."

Purpose and Outline of This Book 2

Abstract

Reviewing the double-slit experiment as a starter, we raise very basic questions we challenge:

1. The roles of the Schrödinger equation and Schrödinger function
2. The pathways of each electron through slit(s)
3. Mechanism of generation of the interference fringe pattern
4. Does the wavepacket collapse on leaving the individual spots?
5. Existence of dynamical equation for individual electron ejected in the experiment?

We address these questions by establishing a three-pillared structure of the Schrödinger dynamics, and each pillar means

1. Derivation of the real-valued Schrödinger and the Schrödinger vector from scratch,
2. Giving the real-valued quantum path integral having an integral measure,
3. Formulating stochastic differential equation for individual quantum path dynamics.

A systematic interpretation of quantum mechanics follows.

Keywords

Double-slit experiment · Wavepacket collapse · Interference pattern · Particle–wave duality · Real-valued Schrödinger equation · Real-valued path integral · Quantum stochastic differential equation

K. Takatsuka, *Structure of the Schrödinger Dynamics*,
https://doi.org/10.1007/978-981-92-1910-0_2

2.1 Double-Slit Experiment: Issues Raised by Experimental Facts

Quantum mechanics is one of the fundamental laws governing natural phenomena in regimes where classical mechanics breaks down (more precisely, where classical mechanics lies outside its domain of applicability). In particular, the Schrödinger equation provides the most important theoretical tool in fields such as atomic and molecular physics and condensed matter science. Accordingly, based on the accumulation of applied research over the past hundred years, no fundamental errors have been found as long as the phenomena under consideration lie within its range of applicability, and the theory itself is therefore believed to be correct.

At the same time, there exists a well-known experiment—the double-slit experiment—that can emphasize, in an extremely concise manner, both the "mystery" and the "limitations" of the Schrödinger equation. Throughout this book, we consistently examine the strangeness and the limits of quantum theory by taking as examples the phenomena that appear in the double-slit experiment.

Conceptually, the double-slit experiment has the following configuration (see Fig. 2.1). The experiment is performed in a vacuum and consists of three elements:

1. A gun capable of emitting particles like electrons one by one
2. A planar detector that receives the emitted particles and senses their arrival positions

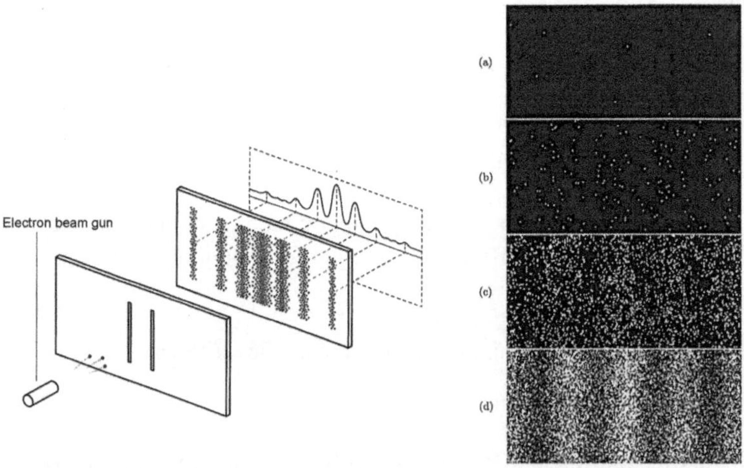

Fig. 2.1 Schematic diagram of the double-slit experiment. Left panel: Schematic of the experimental setup. Right panel: Experimental results by Tonomura et al. The trace of each individual electron arriving at the detector screen. (Adapted from Tonomura, "Beautiful Quantum Mechanics Seen with the Eyes" (Science-sha), with permission)

3. A shielding plate, placed at an appropriate position between the gun and the detector, which has two or more (usually two) independent channels (slits) through which the emitted particles may pass

Particles that do not pass through any channel are excluded and do not reach the detector. A "classical particle" can pass through only a single channel and, after passing through it, reaches the detector and is captured. The experiment is repeated many times, and the positions of the captured particles are accumulated and displayed as bright spots.

Initially, the experiment was performed using photons. Later, however, Tonomura (Akira Tonomura) and coworkers successfully replaced the emitted particles with electrons and carried out an analogous experiment, thereby achieving a de Broglie-type revolution [1]. This was accomplished in 1989 and is comparable in importance to the electron diffraction experiments conducted successively between 1927 and 1928 by Davisson, Germer, and Kikuchi (Masashi Kikuchi), which will be discussed later. In the experiments by Tonomura and colleagues, the detector consisted of a substrate uniformly coated with molecules that emit light upon recombination with electrons; the emitted light was amplified to reproduce the arrival positions of electrons on the detection screen. Because an electron gun can emit electrons one by one, the "particle picture" of electrons is directly employed, and since a spot is recorded on the detector upon the arrival of each electron, the particle nature is directly observed. However, the form taken by the electron during its propagation—including its passage through the channels from emission to detection—remains unknown. (If one were to devise an apparatus to make this known, the original purpose of the experiment would be destroyed.)

Figure 2.1 shows how the spot pattern left by arriving electrons accumulates over time. Shortly after the start of the experiment, when the number of emitted electrons is small, the spots appear to be detected almost randomly (Fig. 2.1, panels a and b). When the number of detected electrons exceeds about 6000, a striped distribution begins to emerge (panel c), and with 40,000 electrons, a clear fringe pattern appears (panel d).

This fringe pattern disappears if either one of the two channels is closed. Moreover, even if single-channel experiments are performed for each slit separately and the detected spots are subsequently added together, the fringe pattern does not reappear. Thus, the pattern appears as if a single wave had passed simultaneously through both slits and then interfered behind them. For this reason, the pattern is widely referred to as an interference fringe. In addition, even when the channel is restricted to one, the distribution of electrons passing through it spreads more widely than would be expected from the incident angle, giving the appearance of a diffraction phenomenon.

The formation of interference fringes is a highlight of quantum theory. Feynman is said to have remarked that "no one understands this mechanism" and also to have stated that quantum theory can describe phenomena but cannot explain them. The existence of facts that cannot be elucidated despite every scientific effort is both bewildering and fascinating.

On the other hand, one sometimes encounters statements bordering on resignation or giving up (and, from a scientific standpoint, almost on despair), such as "An electron that was certainly emitted as a single particle from the electron gun must have become a wave, passed simultaneously through two channels, interfered with itself, and then returned to being a single particle upon reaching the detector."

It should be added that when experiments are performed to determine which channel the particle passed through, different outcomes occur—for example, the disappearance of the interference fringes. There exist numerous papers and review articles on this subject, to which the reader is referred. Another important point is that interference fringes have also been observed for large molecules such as fullerenes (C_{60}), which are aggregates of many atoms [2]. This is significant because it suggests that the occurrence of interference fringes is not an attribute unique to electrons alone. This, too, is one of the important themes of this book.

The interpretation that "an electron passes through the channels as a wave and undergoes an instantaneous collapse of the wave packet when it returns to being a particle at the detector" was advocated at an early stage by the Copenhagen school and others and has become widely accepted (it is often said that Heisenberg proposed it). However, the sudden collapse of the wave packet is an idea that cannot be derived at all as a phenomenon from orthodox quantum theory, including the Schrödinger equation. Moreover, if the collapse were truly instantaneous, it would violate special relativity, and naturally, no one has ever observed such a process. Is it not possible to establish quantum theory without relying on such an implausible hypothesis?

In fact, extended formulations of quantum theory have been proposed, which do not require wave packet collapse. For instance, the many-worlds interpretation allegedly are now as popular as the Copenhagen interpretation. However, these newly devised theories often introduce unprovable ideas, such as a "wavefunction of the universe" or "branching of worlds," thereby importing new kinds of implausibility and difficulty that can further confuse and complicate understanding.

Finally, let us return to panels a–c of Fig. 2.1. The fringe distribution shown in panel d is reproducible by solving the Schrödinger equation and taking the square of the absolute value of the solution. This may well be called a miraculous success of quantum theory and in particular of the Schrödinger equation. It means that an experiment repeated many times (in principle, infinitely many times) can be represented by a single Schrödinger wavefunction (more precisely, by the square of its absolute value). One is struck by a sensation as if the law of large numbers in dice throwing were operating behind the dynamics.

Then, how should the individual motions of each electron, accumulated one by one in panels a–c of Fig. 2.1, be described? If such a law of motion exists, it cannot be the Schrödinger equation itself, yet it must be consistent with and not contradict the Schrödinger equation. At present, there appears to be no theory that has gained consensus, nor any mathematical theory verified experimentally, concerning such a law of motion. Searching for it is also one of the objectives of this book.

2.2 An Overview and Methodology of Schrödinger Dynamics

In this book, we seek to reconsider the difficulties symbolized by the double-slit experiment by returning to the foundations of Schrödinger dynamics. To that end, we briefly summarize below what we aim to achieve and what we propose as new theories, methodologies, and interpretations.

2.2.1 A Universal Rebuilding of the Schrödinger Equation and Derivation of the Quantum Stochastic Differential Equation for the Individual Dynamics

2.2.1.1 Rebuilding of the Real-Valued Schrödinger Equation

First, we derive the Schrödinger equation over the real number field—without using imaginary numbers—from scratch, by a completely universal method. (Strictly speaking, the expression "to derive a fundamental equation" is not logically appropriate; what is meant here is rather "to devise" or "to construct" such an equation. Whether an equation is "correct" can only be verified by experiment.)

Let $\boldsymbol{q} = (q_{1x}, q_{1y}, q_{1z}, q_{2x}, q_{2y}, q_{2z}, \cdots)$ be a point in the multidimensional configuration space of a system of particles, and let m_k denote the mass associated with the k-th coordinate, t the time, and $\hbar = h/2\pi$ Planck's constant. In this book, unless otherwise stated, a point in configuration space will simply be denoted by q.

In this book, by imposing a couple of space-time translational invariance condition and a simple variational condition on the flux conservation for the density distribution function $\varrho(q, t)$, for which the square root is taken in the form

$$\varrho(q, t) = \begin{pmatrix} \phi_r(q) & \phi_c(q) \end{pmatrix} \begin{pmatrix} \phi_r(q) \\ \phi_c(q) \end{pmatrix}, \tag{2.2.1}$$

we derive the real-valued vector Schrödinger equation for $\begin{pmatrix} \phi_r(q) \\ \phi_c(q) \end{pmatrix}$ such that

$$\hbar \mathbf{J} \frac{\partial}{\partial t} \begin{pmatrix} \phi_r(q) \\ \phi_c(q) \end{pmatrix} = \left(-\frac{\hbar^2}{2m} \nabla^2 + V(q) \right) \begin{pmatrix} \phi_r(q) \\ \phi_c(q) \end{pmatrix} \tag{2.2.2}$$

without invoking any existing physical laws, such as classical mechanics or geometrical optics. Here, $V(q)$ indicates a potential function, and $\mathbf{J}$ denotes the following symplectic standard matrix:

$$\mathbf{J} = \begin{pmatrix} 0 & -1 \\ 1 & 0 \end{pmatrix}. \tag{2.2.3}$$

The vector $\begin{pmatrix} \phi_r \\ \phi_c \end{pmatrix}$ is called the Schrödinger vector. Equation (2.2.2) is exactly equivalent to the original complex-valued Schrödinger equation:

$$i\hbar\frac{\partial}{\partial t}\psi(q,t) = \left(-\frac{\hbar^2}{2m}\nabla^2 + V\right)\psi(q,t) \tag{2.2.4}$$

and the Schrödinger function.

$$\psi(q,t) = \phi_r(q,t) + i\phi_c(q,t). \tag{2.2.5}$$

Incidentally, the Hamiltonian and momentum operators working on the Schrödinger vectors are

$$\hat{H} = \mathbf{J}\hbar\frac{\partial}{\partial t} \quad \text{and} \quad \hat{p} = -\mathbf{J}\hbar\vec{\nabla}. \tag{2.2.6}$$

Since the Schrödinger equation is a fundamental equation, it cannot, of course, be derived from any other theory as aforementioned. However, although its form is simple, the Schrödinger equation is a theory that can be coherently constructed from scratch, and its logical structure shows that this is by no means mysterious. Through this, one can deepen the understanding and interpretation of the physical meaning of the Schrödinger equation and of the wavefunction.

2.2.1.2 Real-Valued Path Integral as the Green Function of the Real-Valued Schrödinger Equation

One can write down the Green function of the real-valued Schrödinger equation, Eq. (2.2.2), in the form of the path integrals in a matrix form as follows:

$$\mathbf{G}(q,t:0,0) = \lim_{N\to\infty}\int_{-\infty}^{\infty}\Delta q_1\cdots\Delta q_{N-1}\left(\frac{m}{2\pi\hbar\mathbf{J}\Delta t}\right)^{(N-1)/2}$$
$$\times \exp\left[\sum_{k=1}^{N-1}\frac{1}{\hbar}\mathbf{J}\left(\frac{m}{2}\frac{(\Delta q_k)^2}{\Delta t} - V(q_k,t_k)\Delta t\right)\right]. \tag{2.2.7}$$

As seen, it possesses a Wiener-type integration measure, in a marked contrast to the Feynman path integrals such that

$$d\mathbf{P}_{W[q(s+\Delta s,\omega),s+\Delta s:q(s,\omega),s]}(\omega) = dq\left(\frac{m}{2\pi\hbar\mathbf{J}\Delta s}\right)^{(N-1)/2}$$
$$\times \exp\left[\Delta s\mathbf{J}^T\left(-\frac{m}{2\hbar}\left(\frac{q(s+\Delta s,\omega)-q(s,\omega)}{\Delta s}\right)^2\right)\right]. \tag{2.2.8}$$

Since Feynman's path integral does not have a well-defined integration measure, its absolute convergence is not guaranteed (indeed, it can easily be proven that it does not absolutely converge). By comparing the path integral in Eq. (2.2.7) with the Feynman-Kac formula—an established path integral in statistical mechanics and the theory of stochastic processes—one can obtain a solid interpretation of the principle of superposition in quantum theory and of its stochastic-process nature.

2.2.1.3 Stochastic Differential Equations as Equations of Motion for the Quantum Dynamics of Individual Events

As seen in Fig. 2.1, panels a–c, of the "double-slit experiment," electrons arrive one by one at individual locations. Only when all these events are summed, an interference pattern appears, as shown in panel d, which is reproduced by the squared absolute value of the Schrödinger wavefunction. Why phenomena arising from a large number of events (in this case, many individual emissions of electrons) can be described by a single Schrödinger wavefunction and why its squared modulus is required for comparison with observations (Born's probability interpretation) are central issues in the interpretation of quantum theory.

An even more intriguing question, however, is how the individual events accumulated in Fig. 2.1, panels a–c, should be described. Does a theory that describes them exist? In this book, we derive a stochastic differential equation for quantum stochastic path dynamics, which describes individual events. It has the same form as the Ito stochastic differential equation for the diffusion equation. Our quantum path stochastic differential equation is

$$dX_t = -\frac{\hbar}{\rho m}\left(\phi_r\ \phi_c\right)\mathbf{J}\nabla\begin{pmatrix}\phi_r\\ \phi_c\end{pmatrix}dt + \sqrt{\frac{\hbar}{m}}dW_0\left(t,\omega\right), \tag{2.2.9}$$

where X_t represents the trajectory of motion of an individual event and $W_0\left(t, \omega\right)$ is a standard Wiener process. $\begin{pmatrix}\phi_r\\ \phi_c\end{pmatrix}$ is the Schrödinger vector defined as above. That is, mathematically, X_t evolves while being driven by the velocity field generated by the wavefunction, and at the same time, it possesses stochasticity corresponding to Brownian motion with a diffusion constant of magnitude $\hbar/2m$.

In this way, the real Schrödinger equation, the real path integral, and the stochastic differential equation of quantum stochastic path dynamics constitute the three pillars of the Schrödinger dynamics to be constructed in this book. Figure 2.2 schematically illustrates these three pillars and further suggests their relationship to the three pillars of diffusion theory.

2.2.2 Interpretation of Quantum Theory Based on an Overview of Schrödinger Dynamics

After presenting the above dynamical structure, we develop the phenomena that are naturally anticipated from it, as well as interpretations and understandings of quantum theory. Most of the historical interpretations of quantum theory that are

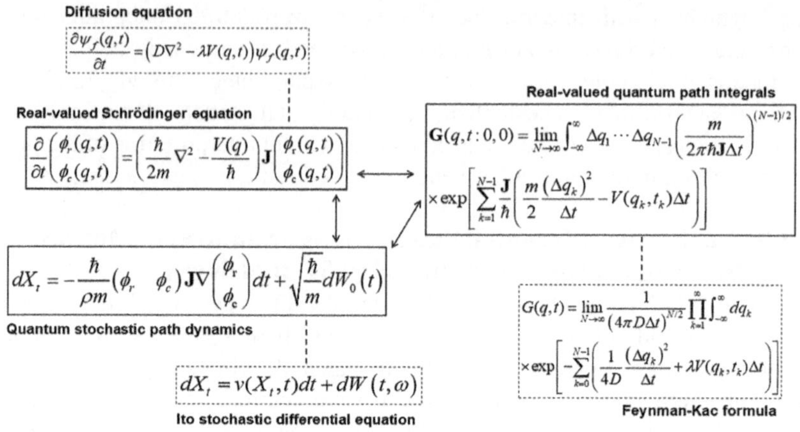

Fig. 2.2 The structure of the three pillars of Schrödinger dynamics. Equations in the solid line boxes represent quantum dynamics proposed in this book, while those in the broken-line boxes are their conterparts in diffusion theory

currently mainstream are based on a monistic view centered on the wavefunction and the Schrödinger equation. Typical examples include the Copenhagen interpretation, the many-worlds interpretation, wave function collapse and related concepts, discussions of peculiar long-range correlations based on the assumption of quantum entanglement, and Bohm's trajectory interpretation.

However, it is essential to distinguish between the entirety of Schrödinger dynamics and the Schrödinger equation itself. Stated more positively, the Schrödinger equation does not constitute the whole of Schrödinger dynamics. The dynamics of individual phenomena (quantum path stochastic differential equations) and the Schrödinger equation as their totality stand in a part-whole relationship. As long as the observation of quantum phenomena is essentially the aggregate of observations of individual events, it is sometimes necessary to distinguish them clearly. In other words, the theoretical structure possesses a hierarchical organization of "parts" and "whole." From the stochastic differential equations that describe the quantum dynamics of individual events, a variety of physical consequences emerge.

In this book, from the perspective of an overall view of Schrödinger dynamics that covers its hierarchical structure, we examine the following concepts and interpretations from a fundamental level:

- What a wavefunction is in the first place
- The meaning and role of the Schrödinger equation (in particular, the significance of the appearance of complex numbers)
- Whether the wavefunction collapses
- Whether the wavefunction represents the so-called matter wave

- Whether the de Broglie matter wave is an intrinsic property of matter
- What de Broglie's particle-wave duality is
- The origin of Born's probability interpretation
- Real-valued path integral as Green's functions of the Schrödinger equation and their meaning
- The consistency between statistical additivity and the principle of superposition in quantum theory
- Quantum dynamical equation for the individual physical event
- Why electrons emitted one by one to form interference fringes after all in the double-slit experiment (the relationship between parts and whole)
- The relationship between quantum dynamics in the quantum canonical equations of motion and classical dynamics (Hamilton's canonical equations of motion and Newtonian equation)
- The kinematical origin of the uncertainty relations
- The preservation and breaking of symmetries in the Schrödinger equation
- Quantum entanglement as a "spooky" long-range correlation and spontaneous disentanglement (detanglement), etc.

References

1. Tonomura, A., Endo, J., Matsuda, T., Kawasaki, T. & Ezawa, H. Demonstration of single-electron buildup of an interference pattern. *Am. J. Phys* **57**, 117–120 (1989).
2. Nairz, O., Arndt, M. & Zeilinger, A. Quantum interference experiments with large molecules. *American Journal of Physics* **71**, 319–325 (2003).

Schrödinger Equation: The Location of the Problem

3

Abstract

Before penetrating into the new structure of quantum mechanics, we first revisit the classic and standard scheme of the Schrödinger dynamics and some of the standard interpretation of quantum mechanics as well. The relation of quantum mechanics to classical counterparts such as the Hamilton-Jacobi equation, Liouville equation, Poisson bracket, and so on is first reviewed. This chapter serves as a short summary of basic structure of quantum mechanics in a standard textbook level. By doing so, we demonstrate how quantum mechanics is different from classical counterparts.

Then, some of the widely recognized interpretations of quantum theory are presented along with their possible and potential difficulties. Those are Born's probabilistic interpretation, the Copenhagen interpretation including the wavepacket collapse, EPR paradox, Bohm trajectory interpretation, the uncertainty relation, nonlocal correlation and quantum entanglement to Bell's inequality, many-world interpretation, and so on.

Keywords

Hamilton canonical equation · Hamilton-Jacobi equation · Bohmian trajectory · Liouville equation · Wigner phase-space quantum mechanics · Poisson bracket · Copenhagen interpretation · Born's probabilistic interpretation · Uncertainty relations · wavefunction collapse · Many-worlds interpretation · quantum entanglement and Einstein · Bell's inequality

Understanding quantum theory remains elusive; even now, strong opposing opinions have been expressed by prominent researchers for those who have studied quantum theory for a long time. For example, a Nobel laureate in theoretical physics, famous for his work on the unification of weak and electromagnetic forces, Gerald 't Hooft,

K. Takatsuka, *Structure of the Schrödinger Dynamics*,
https://doi.org/10.1007/978-981-92-1910-0_3

asserted in Scientific American (2025) that "Quantum physics is on the wrong track." Nevertheless, in this chapter, I will first assume that the Schrödinger equation and Heisenberg's equation of motion, etc., are themselves established theories that do not contain any fundamental errors. With that premise, I would like to first examine what the issues are and what is hindering understanding. Based on this, from the next chapter onward, I will reconstruct Schrödinger dynamics and develop a deeper understanding and interpretation of quantum theory, aiming to explore quantum worlds beyond the Schrödinger equation.

3.1 Schrödinger Equation

3.1.1 Schrödinger Equation as an Axiomatic Principle

The Schrödinger equation, which suddenly appeared, takes the form:

$$i\hbar\frac{\partial}{\partial t}\psi(\boldsymbol{q},t) = \left(-\sum_k \frac{\hbar^2}{2m_k}\nabla_k^2 + V(\boldsymbol{q},t)\right)\psi(\boldsymbol{q},t). \tag{3.1.1}$$

Hereafter, V generally represents a potential function. In what follows, this equation is simply written as

$$i\hbar\frac{\partial}{\partial t}\psi(q,t) = \left(-\frac{\hbar^2}{2m}\nabla^2 + V\right)\psi(q,t). \tag{3.1.2}$$

The vector like $\boldsymbol{q}$ is simply written as a scalar unless otherwise noted. The wavefunction $\psi(q,t)$ is a complex function and can be expressed as

$$\psi(q,t) = \phi_r(q,t) + i\phi_c(q,t)$$

with two real-valued functions. Equation (3.1.2) is also represented as

$$i\hbar\frac{\partial}{\partial t}\psi = \hat{H}\psi \tag{3.1.3}$$

with the $\hat{H}$ being the Hamiltonian, and the momentum operator is

$$\hat{p} = -i\hbar\vec{\nabla}. \tag{3.1.4}$$

This is claimed to be our fundamental equation, and since the fundamental equation is defined as such, it cannot be derived from other fundamental equations. Therefore, in the world of natural sciences rather than mathematics, it is already recognized as a "axiomatic principle." Yet the question is again what exactly the

Schrödinger equation and the wavefunction are. What do they reflect? Is it really acceptable for the fundamental equation to contain imaginary numbers?

3.1.2 Quantum Mechanics Versus the Schrödinger Equation

Dirac defined the state vector and described quantum phenomena in a linear state space [1]. It is essentially the same as the Hilbert space, thereby expressing the fundamental mathematical structure of quantum mechanics with beauty. The state vector $|\psi(t)\rangle$ (called a ket) forms a space, and the corresponding dual vector $\langle\psi(t)|$ (called a bra) forms another space. Operators like the evolution operators act on these spaces. Between these two spaces, the inner product

$$\langle\phi\,|\psi\,\rangle = \langle\psi\,|\phi\,\rangle^* \tag{3.1.5}$$

is defined. The asterisk $*$ on the right denotes complex conjugation. The wavefunction, written as $\psi(q,t)$ in the component form, can be written as the inner product:

$$\psi(q,t) = \langle q\,|\psi(t)\,\rangle \tag{3.1.6}$$

However, the choice of the coordinate system $\{q\}$ is arbitrary, and since it can also be represented in momentum space

$$\tilde{\psi}(p,t) = \langle p\,|\psi(t)\,\rangle \tag{3.1.7}$$

is also a legitimate wavefunction. These wavefunctions are related to each other by

$$\tilde{\psi}(p,t) = \int dq\,\langle p\,|q\,\rangle\,\langle q\,|\psi(t)\,\rangle \tag{3.1.8}$$

or

$$\psi(q,t) = \int dp\,\langle q\,|p\,\rangle\left\langle p\,\middle|\tilde{\psi}(t)\right\rangle \tag{3.1.9}$$

as can be seen from the transformations above. In this way, various representations of the entirety of quantum mechanics are mutually related through such transformations, forming a grand structure. Incidentally

$$\langle q\,|p\,\rangle = \frac{1}{\sqrt{2\pi\hbar}}\exp\left(\frac{i}{\hbar}pq\right) \tag{3.1.10}$$

and

$$\langle q_1\,|q_2\,\rangle = \delta(q_1 - q_2) \tag{3.1.11}$$

are such elements inducing transformations. This Dirac delta function is also one of his great achievements, and it laid the foundation for the mathematical field known as the theory of generalized functions (Schwartz's distribution and Sato's hyperfunction).

Furthermore, the time evolution of the state vector can be written in the form of Eq. (3.1.3) using the time evolution operator:

$$\hat{U}(t, t_0) = \exp\left(\frac{1}{i\hbar}\hat{H}(t - t_0)\right) \tag{3.1.12}$$

for the time-independent Hamiltonian and can be expressed as

$$|\psi(t)\rangle = \hat{U}(t, t_0)\,|\psi(t_0)\rangle\,. \tag{3.1.13}$$

For the wavefunction

$$\psi(q, t) = \langle q|\,\hat{U}(t, t_0)\,|\psi(t_0)\rangle \tag{3.1.14}$$

holds.

In quantum theory, "physical quantities" are represented by operators, and the physical quantities corresponding to the numerical values we recognize in classical theory are expressed as the expectation values of operators. For example, the expectation value of an operator $\hat{A}$ is represented as

$$\left\langle \hat{A}(t) \right\rangle = \langle\psi(t)|\,\hat{A}\,|\psi(t)\rangle\,. \tag{3.1.15}$$

However, this definition is not self-evident, even if we consider it as corresponding to the average value (expectation value) in probability theory. Moreover, the operator corresponding to a physical quantity that can be observed classically must be self-adjoint. In other words, it must be Hermitian in that

$$\left\langle \psi_1 \left| \hat{A}\psi_2 \right.\right\rangle = \left\langle \hat{A}^\dagger \psi_1 \left| \psi_2 \right.\right\rangle, \tag{3.1.16}$$

and it is necessary that the domain of definition $D(\hat{A})$ of $\hat{A}$ and the domain of definition $D(\hat{A}^\dagger)$ of the adjoint operator $\hat{A}^\dagger$ coincide. The state vector that satisfies the eigenvalue equation

$$\hat{A}\,\left|\phi_{a_i}\right\rangle = a_i\,\left|\phi_{a_i}\right\rangle \tag{3.1.17}$$

is called an eigenvector belonging to the real eigenvalue a_i ($i = 1, 2, \cdots$) and plays an extremely important role. The operator $\hat{A}$ is represented in the eigenvalues and their eigenfunctions as

$$\hat{A} = \sum_i |\phi_{a_i}\rangle a_i \langle\phi_{a_i}| \,. \tag{3.1.18}$$

In case of the presence of a set of orthonormalized basis functions $\{\phi_k \,| k = 1, 2, \cdots, N\}$, the resolution of unity (operator) [2]

$$\hat{I} = \sum_k^N |\phi_k\rangle \langle\phi_k| \tag{3.1.19}$$

works as a very convenient element. For instance, if this set is complete, covering all the relevant functional space, a product of two arbitrary operators $\hat{A}$ and $\hat{B}$, we may write

$$\hat{A}\hat{B} = \hat{A}\hat{I}\hat{B} = \sum_k \hat{A} |\phi_k\rangle \langle\phi_k| \hat{B}. \tag{3.1.20}$$

Equation (3.1.18) is merely an example of

$$\hat{A} = \hat{I}\hat{A}\hat{I}. \tag{3.1.21}$$

Even when the basis set covers only a finite subspace, $\hat{I}$ serves as a projection operator being idempotent such that

$$\hat{I}^2 = \hat{I}. \tag{3.1.22}$$

Likewise, the wavefunctions have been lifted to stated vectors and the components of operators, which is the genuine value of Dirac theory.

Thus, the question "What is a wavefunction?" has, through Dirac's abstraction, been replaced by the more general question "What is a state vector?"—but has the reader been convinced? The author, too, once devoured Dirac's book *The Principles of Quantum Mechanics* (4th edition) [1] with great excitement in his youth but, at the time, did not fully understand.

3.1.3 Basic Properties of the Schrödinger Equation

The Schrödinger equation, when solved with appropriate initial and boundary conditions, yields a unique wavefunction as its solution. In this sense, it is a well-posed equation. Strictly speaking, since it does not possess the Lorentz invariance, it differs from the standpoint of special relativity, but this is a matter of the limits of its applicability and not necessarily a fundamental validity. Whether the equations of natural science are correct is, in principle, verified by comparing them with experimental facts, so mathematical correctness is a separate issue.

The Schrödinger equation, inspired by de Broglie's matter waves, was able to reproduce the spectrum of the hydrogen atom without any prior assumptions. For this reason, Schrödinger himself considered that the wavefunction represents the wave motion of the electron in the hydrogen atom. Later, however, criticisms by Ehrenfest and others led to the abandonment of the idea that "wavefunction = physical wave," thereby making the physical meaning of the wavefunction unclear.

3.1.3.1 Probability Density

Regarding the question of what the wavefunction actually is, it is now widely accepted that the consensus is the probability interpretation. Let us define

$$P(q,t) = \psi^*(q,t)\psi(q,t). \tag{3.1.23}$$

Born claimed in 1926 that $P(q,t)dq$ is proposal, proportional to the "probability" of finding a set of particles in the infinitesimal volume dq. This marks the beginning of the probability interpretation. When studying quantum scattering theory, the Born approximation emerges as the lowest-order approximation in the calculation of scattering amplitudes. In this way, he was a pioneer in quantum scattering theory. In the scattering phenomena of a single particle from a target, we consider the situation where the scattered particle flies off in various directions, so the wavefunction may be broken off and can spread out in various directions. Therefore, it is not appropriate to regard the wavefunction as an inherent property of a particle itself. Thus, he noticed that it is better to interpret $P(q,t)dq$ in Eq. (3.1.23) as the probability that a particle is found in the infinitesimal region dq.

However, the Schrödinger equation itself is not logically tied to the probability interpretation. For example, why we should take the product of two complex conjugates, or even why the fact that the wavefunction must be a complex function, cannot be directly understood from probability interpretation. The Schrödinger equation reproduces the spectrum of hydrogen atoms, but this does not necessarily directly connect to the probability interpretation. From the standpoint of the equations, the probability interpretation is something figured out from the outside. For us, if we cannot help but "interpret" it as probability, then Schrödinger's equation cannot be sufficiently fundamental as a basic equation of natural science, as Einstein felt, and he added in a letter to Born, "God does not play dice."

3.1.3.2 Continuity Equation as Noncompressible Flow

However, there is a powerful background to this Born's interpretation. Consider the Schrödinger equation

$$i\hbar\frac{\partial}{\partial t}\psi(q,t) = \left(-\frac{\hbar^2}{2m}\nabla^2 + V\right)\psi(q,t) \tag{3.1.24}$$

and its complex conjugate

$$-i\hbar\frac{\partial}{\partial t}\psi^*(q,t) = \left(-\frac{\hbar^2}{2m}\nabla^2 + V\right)\psi^*(q,t). \tag{3.1.25}$$

Now, as is apparent, if the potential function V is a real function, then by multiplying $\psi^*(q,t)$ to the first equation and $\psi(q,t)$ to the second and then subtracting the two, with a bit of manipulation, we obtain

$$\begin{aligned} i\frac{\partial}{\partial t}(\psi^*\psi) &= -\frac{\hbar}{2m}(\psi^*\nabla^2\psi - \psi\nabla^2\psi^*) \\ &= -\frac{\hbar}{2m}\vec{\nabla}\cdot(\psi^*\vec{\nabla}\psi - \psi\vec{\nabla}\psi^*), \end{aligned}$$

or

$$\frac{\partial}{\partial t}P(q,t) = -\vec{\nabla}\cdot\frac{\hbar}{2im}(\psi^*\vec{\nabla}\psi - \psi\vec{\nabla}\psi^*). \tag{3.1.26}$$

In this expression

$$\vec{j} = \frac{\hbar}{2im}\left(\psi^*\vec{\nabla}\psi - \psi\vec{\nabla}\psi^*\right) \tag{3.1.27}$$

can be regarded as the flux of the flow, and thus, the following equation

$$\frac{\partial}{\partial t}P(q,t) = -\vec{\nabla}\cdot\vec{j} \tag{3.1.28}$$

means, according to Gauss's theorem, that the dynamics of $P(q,t)$ correspond to those of a noncompressible fluid and that probability conservation holds at every local point in space and time. Schiff calls $\vec{j}$ the probability current density [3]. Conversely, if $P(q,t)$ is a probability density, then it is also appropriate to call $\vec{j}$ the flux. (In this way, in natural science, approaching the fundamental law, the circulation way of definitions and tautology appears more often.) Even so, it is questionable whether it is appropriate to call $\vec{j}$ a flux, which contains the imaginary unit i in its definition ($\vec{j}$ itself takes real value though). In classical fluid mechanics, the velocity field $\vec{v}(q,t)$ of the flow appears as in the form:

$$\vec{j}(q,t) = \vec{v}(q,t)P(q,t). \tag{3.1.29}$$

We will examine later whether this form also holds in quantum mechanics.

The above conservation law is used on the assumption that the potential function V is a real function. If V is a complex function, Eq. (3.1.28) changes to

$$\frac{\partial}{\partial t}P = -\vec{\nabla}\cdot\vec{j} + \frac{1}{i\hbar}(V - V^*)P, \tag{3.1.30}$$

and the probability conservation of P is broken. Such complex potentials are called optical potentials and frequently studied in dissipative quantum systems.

3.1.3.3 Time Reversal Symmetry

Since the complex conjugate of the Schrödinger equation

$$i\hbar \frac{\partial}{\partial(-t)} \psi^*(q,t) = \hat{H}\psi^*(q,t) \tag{3.1.31}$$

can also be read as such, $\psi^*(q,t)$ is regarded as a solution of the time-reversed Schrödinger equation. It simply possesses the symmetry:

$$\psi^*(q,t) = \psi(q,-t). \tag{3.1.32}$$

Therefore, the Born probability density $P(q,t)$ can be written as

$$P(q,t) = \psi(q,-t)\psi(q,t) \tag{3.1.33}$$

Thus, the probability density at the present time t is formed by $\psi(q,t)$, which evolved from the "edge" in the past, and by $\psi(q,-t)$, which evolved backward in time from the "edge" in the future, as can be interpreted. There are also serious arguments that the evolution backward from the future serves as one of the first quantum principles (for example, Time-Symmetric Formulation of Quantum Mechanics by Aharonov) [4–6].

3.1.4 Classic and Celebrated Books on Quantum Mechanics

This book aims to develop what may be regarded as an advanced course in Schrödinger dynamics, beginning in Chap. 5. The preceding chapters are therefore devoted to organizing the standard knowledge of quantum theory and clarifying the associated conceptual issues necessary for this purpose. Accordingly, the exposition of general quantum-mechanical knowledge is kept to the essential minimum. Fortunately, there have been so many excellent and traditional textbooks or monographs on quantum mechanics published that we can rest on them. Some of them are cited here by the names of the authors: Dirac [1], Schiff [3], Messiah [7], Landau and Lifshitz [8], Cohen-Tannoudji [9], Weinberg [10], Sakurai [11], and so on. For more specialized topics, there are, for example, von Neumann (mathematical structure) [2], Feynman (path integrals) [12], Wigner (symmetry and group theory) [13], and Newton (scattering theory) [14]. In recent years, an enormous number of books have been published on group theory and representation theory in relation to quantum theory, physics, and chemistry—far too many to list exhaustively. Although Cohen-Tannoudji's work is not yet available in Japanese, it is nevertheless rare, even by global standards, that so many classics can be read in one's native language. Needless to say, the works of Shinichiro Tomonaga (Quantum Mechanics I (2019) and Quantum Mechanics II (2023), Misuzu Shobo) stand among these distinguished contributions.

All of these works are highly esteemed classics and have played an immense historical role. Their presentations are elegant and systematically organized, and—apart from the still unresolved issues of interpretation in quantum theory—they provide outstanding accounts for understanding the structure and applications of quantum mechanics mainly based on the Schrödinger equation. Over the past century, they have served not only as indispensable references for researchers but also as very basic educational texts for countless students.

These textbooks share, to a large extent, a common structure. First, they briefly discuss the necessity of quantum theory, typically in terms of the breakdown of classical mechanics. Second, the Schrödinger equation is introduced in a somewhat ad hoc manner, and its typical solutions, namely, wavefunctions, are presented along with their properties. At the same time, Born's probability interpretation and the so-called Copenhagen interpretation are described. Third, approximation methods such as perturbation theory and the variational method are developed. Fourth, operator algebra, including the Heisenberg equation, is introduced. Fifth, scattering theory and semiclassical formulations are discussed. Sixth, the theory is applied to atoms, molecules, and related systems. Finally, further developments include topics such as the interaction between electrons and electromagnetic fields and the relativistic Dirac equation.

Naturally, individual books differ in emphasis, selection of topics, and distinctive features. For example, the work of Landau and Lifshitz [8] contains many discussions—such as those on semiclassical mechanics and molecular electronic states—that are rare elsewhere. The treatment of path integrals by Feynman [12] is, of course, particularly notable. However, it may be said that, at the level of standard textbooks, there are only very few works that directly address the derivation of the Schrödinger equation itself. Indeed, soon after these textbooks typically discuss the breakdown of classical mechanics, they tend to quickly make a brief account of the discovery that momentum and energy correspond to the wave number and frequency of matter waves (with the notable exception of Tomonaga's *Quantum Mechanics I*; see also Yoshitaka Yamamoto's *Reading Quantum Theory through Bohr and Einstein* (Misuzu Shobo)). Thus, the deep origin and interpretation of the Schrödinger equation are treated in a rather cursory manner.

In relatively recent textbooks (e.g., Weinberg [10] and Sakurai [11]), it is common to regard the wavefunction from the outset as a special type of wave packet, derive plane waves through Fourier decomposition, introduce operators for momentum and energy, and then construct the Schrödinger equation by invoking energy conservation (this will be briefly revisited around Eq. (3.2.47) in this chapter). It is also well known that the Schrödinger equation can be formally obtained from the infinitesimal time evolution in Feynman's path integral formulation. However, since the path integral representation is essentially the Green's function of the Schrödinger equation as a differential equation, deriving the equation it satisfies does not necessarily, in itself, constitute a derivation of the Schrödinger equation.

In short, the existence of the Schrödinger equation has traditionally been treated as something to be accepted almost as a given premise. This may be partly due to the almost miraculous circumstances under which Heisenberg and Schrödinger "discovered" the formalism of quantum theory, as well as the intricate and delicate debates between Bohr and Einstein—topics that, while central to the nature of quantum theory, are not always suitable for detailed reproduction in textbooks. Of course, if one adopts the standpoint that the Schrödinger equation, as a fundamental theory, cannot be derived from anything more basic, such an axiomatic approach is justifiable. However, presenting it in an essentially ad hoc manner risks obscuring its physical meaning and its domain of validity and applicability. At the very least, it deprives the reader of a rational pathway to understanding. As a result, a certain vagueness and elusiveness tend to accompany the understanding of quantum theory. A common piece of advice to beginners has been, "Do not worry about interpretation—first become familiar with the formalism," sometimes expressed more bluntly as "Just calculate" [15].

From this perspective, rather than building arguments upon an unclear and fragile foundation, it may in fact be more transparent—and even more economical in terms of both time and textbook volume—to adopt a deliberately axiomatic formulation. The author's skill is then reflected in the choice of axioms or "requirements for theoretical construction" and in how the logical structure is developed. In this regard, *Foundations of Quantum Theory* by Akira Shimizu (Science-sha) is a particularly thorough and excellent example.

For readers interested in reliving the profound confusion that accompanied the historical development of quantum mechanics, notable works include Yoshitaka Yamamoto's aforementioned study, Kiyoshi Amano's *History of Quantum Mechanics, and Einstein's Struggles with Quantum Theory: A Reappraisal* by Home and Whitaker [16]. There are also more narrative-style works conveying a vivid sense of the debates, such as Manjit Kumar's *Quantum: Einstein, Bohr and the Great Debate about the Nature of Reality* [17]. In addition, broader reflective essays on quantum science, such as Fumitaka Sato's *The Ideology of Quantum Mechanics*, are of considerable value.

In contrast, the present book re-examines the structure underlying the Schrödinger equation, identifying both the process by which it emerges and its meaning, while pursuing a theory of dynamics that goes beyond it through detailed and concrete analysis. Furthermore, by explicitly constructing a real-valued path integral representation, the convergence properties of Feynman's path integral are improved, leading to the formulation of stochastic differential equations for individual quantum dynamics. Such quantum stochastic differential equations are not addressed at all in the classic books aforementioned—indeed, they did not exist at the time.

3.2 Schrödinger Dynamics and the Law of Motion

The Schrödinger equation has withstood a century of scrutiny and is regarded as a correct theory. In the natural sciences, what does "correctness" mean? What is the basis or foundation for determining correctness? Classical mechanics is undoubtedly recognized as correct in expressing the law of motion within its applicable domain. The law of motion, in classical theory, refers to the fundamental law that describes how forces act on objects and what kind of motion they induce. On the other hand, the Schrödinger equation is more strongly perceived as a "law describing the evolution of the state" rather than a "law of motion." Therefore, an attempt has been made to tie quantum mechanics to the "correct representation of classical mechanics," simply because a theory that cannot produce those results that converge to the correct results is hard to be correct. I would like to introduce a representative example here, which will become necessary later in this book.

3.2.1 Existence of Local Point of Application of Force: Various Forms in Classical Mechanics

3.2.1.1 Newtonian Equation and the Hamilton Canonical Equation

In the Newtonian equation

$$f = ma, \tag{3.2.1}$$

when a force f is applied to a mass point of mass m, it is directly observed that the velocity changes over time, and this relationship can be expressed in a single equation. Furthermore, by using the potential energy $V(q)$ as in

$$m\frac{d^2q(t)}{dt^2} = -\frac{\partial V(q)}{\partial q}, \tag{3.2.2}$$

the point of application of the force acting on the mass point can be clearly indicated. The point of application of the force acting on a moving object is localized. On the other hand, in the Schrödinger equation, Eq. (3.1.2), the point of application is not visible. Why did this happen? In this book, we will return to this point in Chap. 8. Also, in Chap. 8, we will introduce E. Nelson's stochastic Newton equation [18]. He argued that the Schrödinger equation can be derived from his stochastic Newton equation. In his theory though, the point of force application becomes ambiguous, and it seems to disappear within the Schrödinger equation.

It is well known that as aforementioned, in the Hamiltonian canonical equations of motion, the momenta are defined as independent variables, and Newtonian mechanics is lifted into a broader phase space. This is done using the classical Hamiltonian:

$$H_{cl} = \frac{p^2}{2m} + V(q) \tag{3.2.3}$$

and

$$\frac{d}{dt}\begin{pmatrix} p \\ q \end{pmatrix} = \mathbf{J}\begin{pmatrix} \frac{\partial H_{cl}}{\partial p} \\ \frac{\partial H_{cl}}{\partial q} \end{pmatrix} \tag{3.2.4}$$

as written. **J** is defined by

$$\mathbf{J} = \begin{pmatrix} 0 & -1 \\ 1 & 0 \end{pmatrix} \tag{3.2.5}$$

and is called the symplectic unit matrix (also known as the standard symplectic matrix). It is a matrix that embodies the symplectic structure of classical mechanics [19]. For example, the conservation law of energy is well known to be

$$\begin{aligned} \frac{dH_{cl}}{dt} &= \frac{dq}{dt}\frac{\partial H_{cl}}{\partial q} + \frac{dp}{dt}\frac{\partial H_{cl}}{\partial p} \\ &= \begin{pmatrix} \frac{dq}{dt} & \frac{dp}{dt} \end{pmatrix} \mathbf{J} \begin{pmatrix} \frac{dq}{dt} \\ \frac{dp}{dt} \end{pmatrix} \\ &= 0 \end{aligned} \tag{3.2.6}$$

and the relation in the second row is called skew orthogonality. **J** plays an important role not only in classical mechanics but also in real-value quantum theory, which begins in Section 5.

From the viewpoint of the point of application of force in quantum theory, let us look at Ehrenfest's theorem [3]. Using the expectation values of the velocity and dynamical variables of a particle, we obtain the expectation values of the canonical equations:

$$\frac{d}{dt}\langle q \rangle = \frac{1}{m}\langle p \rangle = \frac{1}{m}\frac{\hbar}{i}\int dq \psi^*(q,t)\frac{\partial}{\partial q}\psi(q,t) \tag{3.2.7}$$

and

$$\frac{d}{dt}\langle \hat{p} \rangle = -\left\langle \frac{\partial V}{\partial q} \right\rangle. \tag{3.2.8}$$

In this case, the point of application is the expectation value according to the wavefunction. However, in these expressions, the left and right sides are not exactly the same in terms of the way to take expectation values, because $\hat{p}$ is an operator. What is "wavefunction" at all that demands us to use such a kind of expectation value of operators?

3.2.1.2 Hamilton-Jacobi Equation

The family of trajectories derived from the Newtonian equations of motion forms a manifold in configuration space, and the set of these is described by the nonlinear partial differential equation known as the Hamilton-Jacobi (HJ) equation:

$$\frac{\partial S}{\partial t} + \frac{1}{2m}(\nabla S)^2 + V = 0. \tag{3.2.9}$$

Conversely, the classical trajectories lie on the surface formed by the solutions of the HJ equation. Although the HJ equation is a partial differential equation and the action point does not appear explicitly, it is actually present along integral line on the surface, and the particle moves through space-time along it. The action function that defines the integral line

$$S = \int pdq - Edt \tag{3.2.10}$$

is a line integral along

$$p = \frac{\partial S}{\partial q}, \qquad E = -\frac{\partial S}{\partial t}. \tag{3.2.11}$$

Attempting to Tie the Schrödinger Equation

Schrödinger is said to have correlated his theory to the stationary-state HJ equation in formulating the energy spectrum of the hydrogen atom. First, we start with the stationary HJ equation:

$$\frac{1}{2m}(\nabla W)^2 + V = E. \tag{3.2.12}$$

By substituting

$$W = K \log \psi \tag{3.2.13}$$

into it, we obtain that

$$\frac{K^2}{2m}\left(\frac{\partial \psi}{\partial q}\right)^2 = (E - V)\,\psi^2. \tag{3.2.14}$$

Furthermore, considering the total spatial integral on the left-hand side and performing partial integration, we get

$$\int_{-\infty}^{\infty} dq \left(\frac{\partial \psi}{\partial q}\right)^2 = \left[\psi \frac{\partial \psi}{\partial q}\right]_{-\infty}^{\infty} - \int_{-\infty}^{\infty} dq \psi \frac{\partial^2 \psi}{\partial q^2}.$$

Furthermore, if we assume the following boundary conditions

$$\psi(\pm\infty) = 0 \tag{3.2.15}$$

or

$$\frac{\partial\psi(\pm\infty)}{\partial q} = 0, \tag{3.2.16}$$

we have

$$\int_{-\infty}^{\infty} dq\psi\left[-\frac{K^2}{2m}\frac{\partial^2\psi}{\partial q^2} + (V - E)\,\psi\right] = 0, \tag{3.2.17}$$

and we simply choose

$$-\frac{K^2}{2m}\frac{\partial^2\psi}{\partial q^2} + V\psi = E\psi \tag{3.2.18}$$

in the entire space covered by $\psi(q)$. The existing experimental facts naturally impose the constant K as

$$K \to \hbar. \tag{3.2.19}$$

This method of substitution is quite suggestive in considering the meaning of the wavefunction. However, there are still many issues. First of all, the S that appears in Eq. (3.2.9) is assumed to be a real function, but in Eq. (3.2.13), it is extended to the complex number field. Since the analytical properties of the HJ equation are not fully understood, it is unclear whether this substitution (analytic continuation) makes sense.

Furthermore, the boundary conditions in Eqs. (3.2.15) and (3.2.16) are not necessarily appropriate for the wavefunction in a stationary state. In the theory of stationary state scattering, the most essential information is the wavefunction and flux at infinitely remote sites, far from the scattering region, and it is not expected that they should become identically zero in the asymptotic region. The most important physical quantity in the scattering process, the scattering cross section, is also based on the scattering amplitude calculated from the flux. Therefore, the boundary conditions in Eq. (3.2.15) and/or (3.2.16) should not be imposed to establish Eq. (3.2.18).

Bohm's Representation of the Schrödinger Equation

In later years, Bohm took the opposite approach [20]. Based on the premise that the Schrödinger equation is correct, he rewrote the wavefunction as follows:

$$\psi\,(q,t) = R(q,t)\exp\left(\frac{i}{\hbar}S_B\,(q,t)\right). \tag{3.2.20}$$

Here, both $R(q,t)$ and $S_B(q,t)$ are taken to be real-valued functions. Doing so, one obtains an equation similar to the HJ equation for S_B:

$$\frac{\partial S_B}{\partial t} + \frac{1}{2m}(\nabla S_B)^2 + V - \frac{\hbar^2}{2m}\frac{\nabla^2 R}{R} = 0 \tag{3.2.21}$$

and another one for R

$$\frac{\partial R}{\partial t} + \frac{1}{2m}[R\nabla^2 S_B + 2\nabla S_B \cdot \nabla R] = 0. \tag{3.2.22}$$

With the definition of

$$\rho(q,t) \equiv R(q,t)^2, \tag{3.2.23}$$

Equation (3.2.22) yields the following equation of continuity:

$$\frac{\partial}{\partial t}\rho(q,t) = -\nabla \cdot \vec{j}_B(q,t). \tag{3.2.24}$$

This made it clear that a flux similar to that in classical theory appears here such that

$$\vec{j}_B(q,t) = v_B(q,t)\rho(q,t) = \frac{1}{m}(\nabla S_B)\,\rho(q,t). \tag{3.2.25}$$

From this set of equations, the difference between quantum mechanics and classical mechanics is concentrated in the term called the quantum potential in Eq. (3.2.21), which is

$$V_Q(q,t) = -\frac{\hbar^2}{2m}\frac{\nabla^2 R(q,t)}{R(q,t)}. \tag{3.2.26}$$

Furthermore, $V_Q(q,t)$ is a nonlinear function with respect to $R(q,t)$, and the value of $V_Q(q,t)$ at point q appears to indirectly require knowledge of $R(q,t)$ in other regions. That is, one Bohmian trajectory demands the information from other trajectories, and in this sense, they interact with one another in an indirect manner.

It is frequently claimed that Bohm's theory is nothing more than a rewriting of the Schrödinger equation, so in that sense, nothing beyond the Schrödinger equation will be revealed. His theory is now widely known to provide "trajectory interpretation of quantum mechanics," since quantum momentum $p_B(q,t) = \nabla S_B$ along with the position q seem to give the so-called "hidden variables." Afterward, Bohm devoted his life to seeking the physical origin and meaning of the quantum potential $V_Q(q,t)$.

Quantum Effects on Kinematics and Quantum Diffusion

Bohm, by expressing the wavefunction in the form of Eq. (3.2.20), extracted the essence of quantum effects as the quantum potential in the form of Eq. (3.2.26). The price lies in its nonlinearity. Now that the action functional connects classical and quantum mechanics, then what happens if we take the wavefunction in the form

$$\psi(q,t) = F(q,t)\exp\left(\frac{i}{\hbar}S_{cl}(q,t)\right) \tag{3.2.27}$$

as above [21]? Here, $S_{cl}(q,t)$ is the classical action function, which satisfies the HJ equation, that is

$$\frac{\partial S_{cl}}{\partial t} + \frac{1}{2m}(\nabla S_{cl})^2 + V = 0. \tag{3.2.28}$$

Thus, the classical mechanical component is concentrated in $S_{cl}(q,t)$, and all quantum effects are contained in the function $F(q,t)$. $F(q,t)$ is a complex satisfying the following linear equation of motion:

$$\frac{\partial F(q,t)}{\partial t} = \frac{1}{m}\left(-p\cdot\nabla - \frac{1}{2}(\nabla\cdot p)\right)F(q,t) + \frac{i\hbar}{2m}\nabla^2 F(q,t). \tag{3.2.29}$$

We call $F(q,t)$ the action decomposed function (ADF) [22–26]. From Eq. (3.2.29), potential terms disappear, resulting in an equation for kinematics. The effects of dynamics are considered through $p = \nabla S_{cl}(q,t)$.

The reason why only kinematics remains is that, in quantum theory, the kinetic energy is represented by the Laplacian ∇^2. If we look back at

$$\nabla^2\left[F(q,t)\exp\left(\frac{i}{\hbar}S_{cl}(q,t)\right)\right]$$

$$= [F\left(\frac{i}{\hbar}\right)^2(\nabla S_{cl}(q,t))^2 \tag{3.2.30}$$

$$+ \frac{i}{\hbar}F\nabla\cdot\nabla S_{cl}(q,t) + 2\frac{i}{\hbar}(\nabla S_{cl}(q,t))\cdot\nabla F \tag{3.2.31}$$

$$+ \nabla^2 F(q,t)]\exp\left(\frac{i}{\hbar}S_{cl}(q,t)\right), \tag{3.2.32}$$

we see that the term on the right-hand side of Eq. (3.2.30) is absorbed into the HJ equation, and the last term (3.2.32) appears as the purely quantum mechanical term. The second and third terms (3.2.31) can be seen as expressing the coupling between classical and quantum dynamics. These terms reflect the change (including inversion) in the size and shape of an infinitesimal volume in phase space when projected onto q-space along the classical phase space flow (phase flow) [25, 26].

The last term (3.2.32) represents the quantum diffusion that occurs along the classical phase flow. Comparing it with the last term on the right-hand side of Eq. (3.2.29), we see that the diffusion coefficient is purely imaginary, but its absolute value is

$$D = \frac{\hbar}{2m}, \tag{3.2.33}$$

which is shown to coincide with the diffusion coefficient in the quantum path stochastic differential equation in Chap. 8.

Equation (3.2.29) allows for various semiclassical approximate methods, as discussed in [24–26], but these are beyond the scope of this book.

3.2.1.3 Liouville Equation

Next, in order to gain a better understanding of the wavefunction, we would like to consider the Liouville equation. The phase space distribution function $\Gamma(q, p, t)$, like the Schrödinger equation, satisfies a linear partial differential equation, that is

$$\begin{aligned}
&\left[\frac{\partial}{\partial t} + \frac{dq}{dt}\frac{\partial}{\partial q} + \frac{dp}{dt}\frac{\partial}{\partial p}\right]\Gamma(q, p, t) \\
&= \left[\frac{\partial}{\partial t} + \frac{\partial H_{cl}}{\partial p}\frac{\partial}{\partial q} - \frac{\partial H_{cl}}{\partial q}\frac{\partial}{\partial p}\right]\Gamma(q, p, t) \\
&= \left[\frac{\partial}{\partial t} + \frac{p}{m}\frac{\partial}{\partial q} - \frac{dV}{dq}\frac{\partial}{\partial p}\right]\Gamma(q, p, t) = 0.
\end{aligned} \tag{3.2.34}$$

Here, the position q and momentum p are independent variables, but the trajectory $(q(t), p(t))$ can be obtained as a local solution using the canonical equations of motion. The solution curve $(q(t), p(t))$ does not cross others in the (q, p, t) space (the extended phase space). This fact is essential for $\Gamma(q, p, t)$ to be a well-defined distribution function (or for the initial value problem to be well posed), and as discussed in Chap. 9, this kind of non-crossing property plays a fundamental role when considering the concepts of distribution and trajectory in quantum theory as well. Furthermore, $\Gamma(q, p, t)$ preserves volume along the phase flow (the flow generated by the set of trajectories) in phase space. That is

$$\int_{\Omega(t)} \Gamma(q, p, t)dqdp = \text{constant} \tag{3.2.35}$$

can be established, and this itself is one of the Poincaré-Cartan absolute invariances [19]. In this theorem, instead of considering individual classical trajectories, we consider the "entire set of trajectories" contained in $\Omega(t)$.

Here, for the sake of clarity, let us confirm the meaning of the distribution function. $\Gamma(q(t), p(t), t)$, with $\Gamma(q(0), p(0), 0)$ as the initial condition, does not simply correspond to the trajectory set $(q(0), p(0)) \rightarrow (q(t), p(t))$ as a "simultaneous" dynamics, nor does it mean that each trajectory describes the phenomena occurring at the same time. Each trajectory is independent and does not constitute an ensemble of events, even if started from any given time; rather, under the initial conditions at each initial time, the distribution function $\Gamma(q, p, t)$ simply records the individually reaching $(q(t), p(t))$ for an elapsed time t. Later, we will see that the Schrödinger wavefunction describes many events (phenomena) simultaneously and in parallel but not simultaneously, and it is important to correctly recognize how the quantum distribution and classical counterpart are mutually similar and different.

Furthermore, it is interesting to note that quantum mechanics can also be considered in phase space. This is the phase space quantum theory of Wigner [27–30]. First, the phase space distribution function $\Gamma_W(q, p, t)$ is defined by

$$\Gamma_W(q, p, t) = \frac{1}{2\pi\hbar}\int dq \exp(ipx/\hbar)\,\psi^*(q + x/2, t)\,\psi(q - x/2, t)\,, \tag{3.2.36}$$

which is followed by

$$\int dp\Gamma_W(q, p, t) = \rho(q, t) \tag{3.2.37}$$

and

$$\int dq\Gamma_W(q, p, t) = \tilde{\rho}(p, t), \tag{3.2.38}$$

where $\rho(q, t)$ and $\tilde{\rho}(p, t)$ are the quantum density functions on the configuration space and its momentum space, respectively. $\Gamma_W(q, p, t)$ is said to be a Fourier transform with respect to the fluctuation of the density. The evolution equation for $\Gamma_W(q, p, t)$ is given by Zachos et al. [30] and Carruthers and Zachariasen [31]

$$\frac{\partial}{\partial t}\Gamma_W(q, p, t) = -\frac{2}{\hbar}H_{cl}(q, p, t)\sin\left(\frac{\hbar}{2}\left(\overleftarrow{\partial}_p\overrightarrow{\partial}_q - \overleftarrow{\partial}_q\overrightarrow{\partial}_p\right)\right)\Gamma_W(q, p, t). \tag{3.2.39}$$

Here, the differential operator $\overleftarrow{\partial}_p$ denotes the partial derivative with respect to p in $H_{cl}(q, p, t)$ on the left, while $\overrightarrow{\partial}_p$ does the partial derivative with respect to p in $\Gamma_W(q, p, t)$ on the right; the same applies for the q derivatives. At first glance, it appears elegant, but in reality, it is an extremely complicated partial differential equation that is not generally solvable. However, the lowest order term in $\hbar$, i.e., the $\hbar^0$ term, yields the classical Liouville equation. Yet the clear local point of application of force becomes unclear in the higher-order terms.

Quantum theory with phase space structure, such as the Wigner function, possesses a certain compatibility with classical mechanics and is widely used for the analysis and approximation of quantum theory [27–30, 32–34].

3.2.1.4 Poisson Bracket

The classical Liouville equation (3.2.34) can be rewritten using the Poisson bracket. For a general classical physical quantity $A(q, p, t)$, taking $\Gamma(q, p, t)$ as an example, its variation can be written in the following form of the Poisson bracket as

$$\{A, H_{cl}\}_P = \frac{\partial A}{\partial q}\frac{\partial H_{cl}}{\partial p} - \frac{\partial H_{cl}}{\partial q}\frac{\partial A}{\partial p}. \tag{3.2.40}$$

The method of transition from classical theory to quantum theory is to consider the operator $\hat{A}$ corresponding to $A(q, p, t)$ and replace the Poisson bracket with the commutator

$$\begin{aligned} \frac{d\hat{A}_H(t)}{dt} &= \frac{\partial \hat{A}_H(t)}{\partial t} + \frac{1}{i\hbar}\left[\hat{A}_H(t), \hat{H}\right] \\ &= \frac{\partial \hat{A}_H(t)}{\partial t} + \frac{1}{i\hbar}\left(\hat{A}_H(t)\hat{H} - \hat{H}\hat{A}_H(t)\right) \end{aligned} \tag{3.2.41}$$

as established by Dirac [1]. Here, $\hat{A}_H(t)$ is the Heisenberg representation of $\hat{A}$:

$$\hat{A}_H(t) = \exp\left(-\frac{1}{i\hbar}\hat{H}t\right)\hat{A}(t)\exp\left(\frac{1}{i\hbar}\hat{H}t\right). \tag{3.2.42}$$

Here again, the point of action of the dynamics is, as in Ehrenfest's theorem, on the expectation value, that is

$$\left\langle \hat{A}_H(t) \right\rangle = \left\langle \psi(0) \left| \exp\left(-\frac{1}{i\hbar}\hat{H}t\right)\hat{A}(t)\exp\left(\frac{1}{i\hbar}\hat{H}t\right) \right| \psi(0) \right\rangle. \tag{3.2.43}$$

The above jump from the Poisson bracket to the commutator is not likely to be a powerful assistance for the physical interpretation of the Schrödinger equation and the wavefunction.

3.2.2 Is the Matter Wave an Intrinsic Property of Matter?

de Broglie, based on the experimental fact that light, which has a wave number k and angular frequency ω, possesses particle-like properties corresponding to momentum p and energy E, asserted that particles with momentum p and energy E should also have wave-like properties characterized by wave number k and angular frequency ω. Thus, the following relations were established:

$$k = p/\hbar, \qquad \omega = E/\hbar, \tag{3.2.44}$$

The wave nature of electrons was confirmed by experiments observing the ring-shaped fringe patterns formed by electrons passing through thin films of crystals, such as those by Davisson, Germer, and Kikuchi, which are similar to diffraction fringes. By observing the pattern (1927–1928), the concept of "matter waves" seemed to be experimentally confirmed.

Meanwhile in 1921, Ramsauer and Townsend independently reported the results of measurements of the scattering cross section (scattering probability) of low-speed electrons, in order to observe the size of rare gas atoms. They found that the cross-section varies depending on the kinetic energy of the electrons involved in the scattering, and in some cases, the cross-section becomes extremely small, to the extent that it appears as if the electrons were passing through the transparent atoms. This phenomenon is known as the Ramsauer-Townsend effect. The Ramsauer-Townsend effect has also been observed for molecules [35–37] and is now regarded as a manifestation of electron wave diffraction. In this respect, it was a discovery that preceded those of Davisson, Germer, and Kikuchi.

Now, let us consider a free particle in a field with a pure (k, ω) of

$$E = \frac{p^2}{2m}. \tag{3.2.45}$$

From the Einstein-Planck relation (the second equation in Eq. (3.2.44)), we have

$$\omega = \frac{1}{2m}k^2. \tag{3.2.46}$$

Then, for a pure plane wave having a set of (k, ω)

$$\begin{aligned}\psi(q,t) &= N\exp(ikq - i\omega t) = N\exp\left[\frac{i}{\hbar}(pq - Et)\right] \\ &= N\exp\left[\frac{i}{\hbar}\left(pq - \frac{p^2}{2m}t\right)\right]\end{aligned} \tag{3.2.47}$$

is found to be a solution to a dynamical equation like the Schrödinger equation. Note that this wave propagates in multidimensional configuration space, but not in the three-dimensional Euclidean space. This was not trivial to the students in the early days right after the birth of quantum theory (Hideki Yukawa wrote, for example, that he and Shinichiro Tomonaga, both a later Nobel laureate, were puzzled in their youth whether $\psi(q, t)$ is a wave in configuration space or a real wave in three dimensions). It is clear that this plane wave solution does not satisfy the boundary conditions imposed by Schrödinger; that is, it does not satisfy either Eq. (3.2.15) or (3.2.16).

By requiring

$$\hat{p}\psi(q,t) = p\psi(q,t) \quad \text{and} \quad \hat{H}\psi(q,t) = E\psi(q,t) \tag{3.2.48}$$

in Eq. (3.2.47), we naturally attain

$$\hat{p} = \frac{\hbar}{i} \frac{\partial}{\partial q} \tag{3.2.49}$$

and

$$\hat{H} = i\hbar \frac{\partial}{\partial t}. \tag{3.2.50}$$

We here converse the logic such that since a free particle of (p, E) should bear a pure set (k, ω), its dynamics can be represented in terms of the functional form like Eq. (3.2.47). Then, we find that

$$i\hbar \frac{\partial}{\partial t} \psi(q, t) = -\frac{\hbar^2}{2m} \psi(q, t) \tag{3.2.51}$$

emerges spontaneously. However, an explanation based solely on plane waves is insufficient; usually, the discussion is refined by decomposing a wave packet into a Fourier series of plane waves, but the essence of the application of the theory is mostly in the case of plane waves.

Incidentally write a Gaussian wavepacket as

$$\psi(q, 0) = (2\pi)^{-1/4} (\Delta q_0)^{-1/2} \exp\left[-\frac{q^2}{4(\Delta q_0)^2}\right] \tag{3.2.52}$$

and evolve it in time by Eq. (3.2.51) as

$$\psi(q, t) = (2\pi)^{-1/4} (\Delta q_0 + \frac{i\hbar t}{2m\Delta q_0})^{-1/2} \exp\left[-\frac{q^2}{4(\Delta q_0)^2 + 2i\hbar t/m}\right]. \tag{3.2.53}$$

Since it spreads spontaneously over time, as seen above, it is difficult to say that this function represents the wave packet expression of a "particle." For this reason, Ehrenfest prompted Schrödinger himself to reconsider the physical meaning of his wavefunction.

A fundamental question arises here: Does the "wave" represented by the function in Eq. (3.2.47) truly correspond to de Broglie's "matter wave"? Looking back, one may ask whether the statement "a quantum particle, such as an electron, is a physical entity that intrinsically possesses wave-like properties, just as it possesses electric charge" is in fact valid. What, then, is meant by the commonly used expression of wave-particle duality: "Wave-like behavior appears in experiments designed to detect waves, while particle-like behavior appears in experiments designed to detect particles"?

An electron can clearly be regarded as a particle in the sense that it carries a unit electric charge and can be counted on that basis. Moreover, as in the double-slit experiment, the fact that it leaves localized spots on a detection screen provides unambiguous evidence of its particle-like nature. On the other hand, it is also an experimental fact that, in typical setups such as the double-slit experiment, the accumulated distribution of detection events—resulting from the repeated (though not simultaneous) involvement of many electrons—exhibits clear patterns that should be described as diffraction and interference.

If one interprets each individual electron as a physically extended wave, one is naturally led to suppose that, in order to recover a particle-like state, the wave packet must undergo a sudden contraction at the point of observation. This notion of wave packet reduction (collapse), often attributed to Heisenberg, remains one of the major unresolved issues in quantum theory even today.

However, no one has ever directly observed a de Broglie wave experimentally. Still less has the "collapse" of an electron wave packet been observed. From the theoretical side, one may question whether it is logically consistent to input information about an electron as a point particle into the Schrödinger equation and obtain, as its solution, a matter wave. Thus, the core of the problem can be formulated as follows: Does an electron possesses wave-like properties as intrinsic physical attribute of individual particle? If so, does the Schrödinger wavefunction directly represent such a wave? In other words, is it justified to conclude that, unless each electron inherently possesses wave-like properties, phenomena such as diffraction and interference fringes could not arise?

It is also conceivable that interference patterns are not direct manifestations of intrinsic properties of electrons, but rather consequences of other principles, such as the laws of motion. Nevertheless, even after a century of quantum theory, simply examining the Schrödinger equation alone as gifted by nature ultimately leads us back to the unresolved question: What, in fact and after all, is the wavefunction?

In this book, we adopt the standpoint that the wave-like nature of the Schrödinger wavefunction does not necessarily imply that individual particles possess physical waves as intrinsic attribute. Instead, what is the Schrödinger function, and through what (mathematical and physical) processes does it emerge? These questions will be examined in a systematic and universal manner starting from Chap. 5. Before doing so, however, we shall first review the standard interpretations of the Schrödinger equation in order to approach the core of the problem as closely as possible.

3.3 Interpretation of the "Descended" Schrödinger Equation

From Chap. 5 onward, this book undertakes a derivation of the Schrödinger equation from first principles and develops an original formulation of quantum theory. In parallel, a new interpretation of Schrödinger dynamics will be presented. Prior to it, in this section, we briefly review conventional interpretations of the Schrödinger equation and closely related theories. It should be emphasized that what follows reflects the perspective of practitioners who routinely employ the Schrödinger equation and is not intended as a comprehensive commentary on quantum theory.

In particular, we do not enter into historical debates concerning the relationship between the Schrödinger equation and the observer, or consciousness and cognition. For example, in quantum Bayesianism, the quantum probabilities introduced by Born are interpreted as Bayesian probabilities, thereby inviting an empirical probabilistic framework. The present work does not possess the mathematical apparatus required either to support or to refute such views. Likewise, unless necessary, we will not discuss cutting-edge No-Go theorems. A vast literature exists on general interpretations of the wavefunction, to which the reader is referred [38–40].

When discussing the interpretation of quantum mechanics, it is important to recognize that there exist several distinct levels of interpretation:

1. The basic level of interpretation, which presupposes the knowledge necessary to apply quantum mechanics in practice. This level is generally regarded as having been established by the Copenhagen school on the basis of the Schrödinger equation. However, it should be noted that being "basic" does not necessarily imply being correct.
2. Interpretations at a level beyond the Schrödinger equation, which attempt to avoid difficulties such as the collapse of the wavefunction. These include, for example, the many-worlds interpretation, which introduces a broader ontological framework. Various No-Go theorems also play a central role in debates at this level. For instance, Bell's theorem (to be discussed later) [41] addresses the nonlocality of quantum states, while the Kochen-Specker theorem [42] demonstrates that it is impossible to assign definite values simultaneously to all physical quantities. Many such results are deeply involved in interpretational controversies at this level.
3. The metascientific level, which extends the discussion to include the role of human observers and theorists and considers the relationship between quantum theory and human society as a whole.

At the stage of development of quantum mechanics, great figures such as Einstein and Schrödinger posed profound paradoxes with the aim of exposing the deep conceptual difficulties of the theory. This was partly because the structure of Schrödinger dynamics (not the Schrödinger equation) had not yet been sufficiently established and partly because the Schrödinger equation was essentially the only theoretical framework available for understanding quantum phenomena. The efforts devoted to resolving these paradoxes have undoubtedly contributed to a deeper understanding and interpretation of quantum theory. Nevertheless, it is said that younger physicists such as Heisenberg and Pauli mocked Einstein as having become an outdated scientist with respect to quantum theory.

However, the paradoxes raised by these pioneers represent serious challenges posed to quantum mechanics, and they demand careful confrontation, as seriously done by Bohr, along with appropriate counterarguments and rigorous justification. The present author adopts the straightforward view that such fundamental paradoxes should be resolvable not by departing from Schrödinger dynamics (again, Schrödinger "dynamics" does not mean the Schrödinger "equation" alone, as will be discussed later on), but rather through a deep understanding of it—one grounded in both mathematical and physical reasoning.

On the other hand, there also exist a considerable number of discussions that propose new interpretations by effectively setting aside these difficulties or concealing them, as it were "under the carpet." In what follows, we shall not enter into the third, the metascientific level, described above. Yet, from the standpoint of clarifying why a new formulation of Schrödinger dynamics is necessary, we shall briefly introduce certain interpretations that may be regarded as part of the standard background knowledge.

3.3.1 Copenhagen Interpretation as a Classic Paradigm

The body of ideas that has been widely accepted by the "quantum mechanics community" is the Copenhagen interpretation. Taking into account the experiences of the early quantum theory developed by Bohr and others, this set of doctrines emerged in the wake of the birth of the new quantum mechanics by Heisenberg, Schrödinger, and their contemporaries, in order to provide a systematic academic understanding of numerous puzzling phenomena and novel experimental findings.

Because the towering figure of Bohr, who was based in Copenhagen, stood at its center, these ideas are collectively referred to as the Copenhagen interpretation. In reality, however, it is composed of the views of many researchers, including Heisenberg and Pauli, among others. Some of these views were not even endorsed by Bohr himself. Consequently, a variety of formulations can be found under the name of the Copenhagen interpretation.

A representative formulation is that given by von Neumann in 1932, which may be summarized as follows:

1. A quantum system obeys the Schrödinger equation when it is not being observed.
2. Matter exhibits wave-particle duality.
3. Measurements involve uncertainty, and the outcomes follow Born's probabilistic rule.
4. This probabilistic rule serves to reconcile (or sublate, aufheben) the wave-particle duality.
5. Observation plays a definite role in a quantum system, leading to the collapse of the wavefunction upon measurement (thereby separating the quantum system from the observer or measuring apparatus).

In *The Oxford Handbook of the History of Quantum Interpretations* (2022) [40], D. Howard presents his own view in the form of a list accompanied by a relatively detailed discussion. In particular, he emphasizes Bohr's "principle of complementarity" and the notion of the "completeness of quantum mechanics." Complementarity suggests that mutually exclusive concepts—such as wave-particle duality, the uncertainty principle, and determinism versus probabilistic interpretation—are all necessary for a full understanding of quantum theory. In this sense, it partly leads into the domain of metascience.

However, one may ask whether these notions—wave-particle duality, the uncertainty principle, and the contrast between determinism and probabilistic interpretation—each possibly rooted in different foundations, can truly be subsumed under a single unifying concept. This question will also be examined in this book.

Adopting the "completeness of quantum mechanics" as an interpretational principle requires a certain degree of boldness. It is a common understanding in science that no physical theory can establish its own infallibility, correctness, unnecessariness of complement, or completeness in a self-referential manner within its own theoretical framework; ultimately, experimental verification is indispensable. In what sense the Schrödinger equation may be regarded as "correct" is also a question that we shall examine in this book.

An interpretation is not the theory itself, but rather an expression of how one understands or views the theory. For this reason, some have referred to the Copenhagen interpretation as the "Copenhagen doctrine" [43]. Moreover, when asked to summarize the Copenhagen interpretation in a single sentence, Mermin famously quipped, "Shut up and calculate!"—only to add, "But I won't shut up." [15].

Be that as it may, despite possible objections, it is appropriate to acknowledge the great contribution of the Copenhagen group: By providing a kind of standard interpretation, they enabled researchers to proceed into the realm of applications without undue hesitation, and through the accumulation of such successful experiences, they promoted the development of quantum theory itself. In what follows, we shall enter more deeply into the domain of interpretation and understanding.

3.3.2 Born's Probability Interpretation and the Wave-Particle Duality

3.3.2.1 Einstein: God Does Not Play Dice

Today, few doubt the validity of the relation that converts the wavefunction into a probability density:

$$P(q,t) = \psi^*(q,t)\psi(q,t). \tag{3.3.1}$$

Its success is largely due to the accumulation of correct predictions under the "shut up and calculate" doctrine. However, deriving this probability interpretation directly from the Schrödinger equation itself is difficult, though consistency with it is demanded.

Einstein, a founder of the physics of stochastic processes, seemed to believe that probabilistic descriptions are valid only for coarse-grained systems and should not enter the fundamental laws of physics. It is likely that he maintained his belief throughout his life.

3.3.2.2 Probability Density and Time Reversal

In Formula (3.3.1), to calculate the probability density $P(q, t)$, the complex conjugate $\psi^*(q, t)$ of the wavefunction, which implies time reversal, appears. Some people may think of this as simply the operation of taking the complex conjugate, while others may see a strong implication of time reversal (as discussed by Aharonov et al. in "Time-symmetric interpretation" [6]). "The probability density at the present time t is expressed by the product of $\psi(q, t)$, which has evolved from the past, and $\psi^*(q, t)$, which has evolved backward from the future." This may seem somewhat strange from the classical common sense.

3.3.2.3 Did the Probability Interpretation Resolve the Wave-Particle Duality?

According to the Dictionary of Physics (revised edition, Baifukan, 1992), "the probability interpretation, by assigning the meaning of probability amplitude to the wavefunction, successfully unifies 'particle-like' and 'wave-like' properties and provides the foundation for the development of wave mechanics as a consistent theoretical framework." However, although this statement appears to reassure us that "matter possesses wave-particle duality, and the wavefunction may therefore be used with confidence," it does not deepen our understanding of the essential nature of this duality. As repeatedly emphasized, the present work aims to directly address the question: Is the matter wave an intrinsic property of matter, and is the Schrödinger wavefunction a physical entity that directly represents it?

3.3.3 Wavefunction Collapse

In the Copenhagen interpretation, it is asserted that measurement or observation induces the collapse of the wave packet, a view that continues to enjoy broad support, including passive acceptance. In the double-slit experiment described in Chap. 2 (Fig. 2.1, panels (a)–(d)), this interpretation holds that each electron leaves an individual spot on the detection screen because the wave packet collapses toward the point of arrival whenever an electron is detected. If the wave packet described by the wavefunction is regarded as a physical entity, one is naturally led to conclude that a spatially extended wave packet should have suddenly contracted to a single point. In this sense, the notion of wave-particle duality seems to obtain a certain interpretation.

3.3.3.1 Does the Wave Packet Collapse?

If a wave packet possessed a physical reality as an actual wave, then an "instantaneous" collapse could, in principle, violate special relativity; it is therefore natural that Einstein opposed such an idea. Moreover, it is unclear what is meant by the instantaneous collapse of a wave packet—its physical substance is not well defined. Through what kind of interaction with an experimental apparatus does the collapse occur? Is it only human-performed experiments that trigger the collapse of the wave packet? At what stage does the collapse take place? Should one even consider processes occurring within the observer's brain (as in the hypothesis of Wigner's friend)?

On the other hand, as a conceivable physical mechanism for collapse, there exists the theory of spontaneous localization proposed by Ghirardi and collaborators (the Ghirardi-Rimini-Weber theory) [44]. Since such topics lie outside the scope of this book (as will be discussed in detail in Chap. 5, where the wavefunction is regarded as a purely mathematical entity and no physical collapse is assumed from the outset), we merely refer the reader to the review by Bassi et al. [45].

Arguments advocating the necessity of collapse often relate directly to the interpretation of Schrödinger dynamics, as follows. Consider a physical observable represented by an operator $\hat{A}$, and suppose that its expectation value is given by

$$a = \langle \psi(t) | \hat{A} | \psi(t) \rangle , \tag{3.3.2}$$

Then, upon measurement of this observable, the quantum rules assign the value a, and the previously unknown wavefunction prior to measurement must (instantaneously) change into the wavefunction corresponding to the value a. The mathematical procedure that expresses this change is sometimes referred to as the projection postulate.

It should be noted carefully that observations in quantum experiments are performed repeatedly (in general, many times), and each time a different outcome is recorded on the measuring apparatus (see panels (a)–(d) of the double-slit experiment discussed in Chap. 2). Remarkably, the Schrödinger wavefunction represents the outcomes of these infinitely many experiments by a single function, yielding the value a in Eq. (3.3.2) as its expectation value (statistical average). This mechanism is profoundly mysterious—indeed, it may seem even more mysterious than the collapse of the wave packet itself. It is therefore not surprising that some researchers might find it more acceptable to admit wave packet collapse.

At present, there exists no Schrödinger equation describing the individual dynamics that yield the numerical outcomes observed in each single measurement. Furthermore, it is not known whether each individual measurement outcome is accompanied by its own "individual wavefunction" (most likely, it is not). Even if such wavefunctions did exist, discussing their collapse would require additional assumptions. Thus, prior to debating collapse, it is more essential to reconsider what the wavefunction fundamentally represents and whether there exists a form of quantum dynamics corresponding to each individual measurement (i.e., each individual event).

3.3.3.2 The Bohmian Trajectory Interpretation

The concept of wave packet collapse has been subject to considerable criticism due to the lack of a clear mathematical theory and experimental support. Consequently, efforts have been made to construct theories that do not require a collapse process. The theory of Bohm, introduced in connection with Eq. (3.2.21), has come to occupy a significant position in the interpretation of quantum mechanics, despite being sometimes dismissed as merely a reformulation of the wavefunction and the Schrödinger equation.

In this interpretation, no wave packet collapse is required. In this book, beginning with Chap. 5, Bohm's theory will frequently be examined in a comparative manner.

3.3.3.3 The Many-Worlds Interpretation

In the so-called "many-worlds interpretation" [46, 47], the collapse of the wave packet is avoided by incorporating not only the observed system but also the observer, the measuring apparatus, and even the entire universe into the wavefunction. The observer obtains a single measurement outcome in accordance with the probabilities assigned by Born's rule.

Since the observer is effectively assigned to each of the many possible events contained within the standard wavefunction interpretation, from the observer's own perspective, there is neither splitting nor collapse of the wavefunction; there is only a single realized event from the outset. Instead, it is assumed that the world branches according to the Born probabilities—probabilities that would otherwise be revealed only through measurement.

Although this approach appears to skillfully sweep the conceptual difficulties of the Schrödinger equation "under the rug," it leads us to the even more difficult notion of a "splitting of worlds." Furthermore, what is meant by the wavefunction of the universe? A probability distribution is not necessarily discrete; it can also be defined over continuous variables. In that case, the number of worlds might increase with each observation in a manner characterized by the cardinality of the continuum.

While this line of thought is worthy of respect as an attempt to evade wave packet collapse, it is difficult to regard it as belonging to the domain of physics. Nevertheless, the many-worlds interpretation appears today to be as popular as the Copenhagen interpretation, partly due to the increased plausibility brought about by developments such as decoherence theory [48–50].

However, 't Hooft has remarked succinctly regarding the many-worlds interpretation: "I don't care" [43]. The author's position is similar.

3.3.4 The Inseparability of the Quantum System and the Observer

As noted in the discussion of Newton's equation (3.2.1), in classical mechanics, observable phenomena can be described using the language and mathematical expressions familiar to the observer, and there was no conceptual discrepancy separating the observer from the observed. The "cut" introduced by the Copenhagen interpretation—namely, the separation of the quantum system from the observer (or measuring apparatus)—constituted a revolutionary conceptual leap.

This reflects two essential features: (1) There exist phenomena that cannot be adequately described using only the language grounded in our everyday experience, as exemplified by wave-particle duality, and (2) the observer inevitably exerts a direct and irreversible influence on the observed quantum system or becomes inseparably involved with it. The latter point is particularly well known as Heisenberg's uncertainty principle.

(For the uncertainty relations, explanations can be found in any standard textbook on quantum mechanics, such as that by Messiah [7]. Additionally, a detailed and modern discussion is available in Shogo Tanimura's "Diversified Uncertainty Relations" (2016), accessible via the Nagoya University Academic Repository: https://nagoya.repo.nii.ac.jp.)

3.3.4.1 Observation Process Directly Intervening in Phenomena: Heisenberg's Uncertainty Principle

Heisenberg considered a thought experiment in which the position and velocity of an electron are determined by irradiating it with gamma rays and observing the scattered radiation. From this analysis, he concluded that (1) the shorter the wavelength of the gamma rays, the more precisely the position of the electron can be determined; however, (2) shorter wavelengths impart a larger disturbance to the electron's momentum, rendering its precise value indeterminate.

Let $\varepsilon(Q)$ denote the measurement error in position, expressed as a standard deviation, and let $\eta(P)$ be the uncontrollable disturbance imparted to the momentum. Heisenberg's result is expressed as

$$\varepsilon(Q)\eta(P) \simeq h. \tag{3.3.3}$$

This formulation not only made explicit the necessity of incorporating the interaction between nature (as the observed system) and the human observer into physical theory, but also led Heisenberg to the further viewpoint that "physical quantities acquire meaning for us precisely because they are measured" or, furthermore, that "physical theories should be formulated solely in terms of observable quantities."

Einstein is said to have sharply reacted to the statement by Pascual Jordan that "observation not only affects the outcome but actually creates it," by remarking, "Do you really believe the Moon is not there when you are not looking at it?" [51]. This book adopts the standpoint that physical phenomena proceed according to Schrödinger dynamics even in the absence of observation (the Moon moves whether or not it is observed). However, in applying the Schrödinger equation, initial and boundary conditions are assumed theoretically based on physical considerations. That is, while such conditions may be inferred from observations, they cannot be uniquely determined by them. This position is grounded in kinematical uncertainty, as will be discussed later.

Meanwhile, Bohr emphasized that quantum theory involves features such as wave-particle duality and uncertainty relations, which reflect the fact that human perception of phenomena in classical way and quantum causality behind are not mutually in opposition but rather stand in a complementary relationship. He thus advocated the construction of the concept of complementarity, which enables a unified and mutually consistent understanding of these aspects.

3.3.4.2 Uncertainty Relations Arising from the Pairs of Non-commuting Operators

It is well known that a unimodal (singly peaked) function with finite width, such as a Gaussian function, has a Fourier transform that is likewise unimodal with finite width. When the original function is sharply localized (i.e., has a small width), its Fourier transform becomes broadly distributed, and vice versa. Thus, a trade-off relation arises between the two widths—namely, an uncertainty relation. This is not a phenomenon unique to quantum mechanics.

Kennard first established such an uncertainty relation for the position Q and momentum P in quantum theory (1927):

$$\sigma(Q)\sigma(P) \geq \frac{\hbar}{2}. \tag{3.3.4}$$

Here, $\sigma(\hat{A})$ denotes the standard deviation of an operator $\hat{A}$, defined as

$$\sigma(A) = \langle\psi(t)| \left(\hat{A} - \langle A\rangle\right)^2 |\psi(t)\rangle^{1/2} \tag{3.3.5}$$

with the expectation value $\langle A\rangle$

$$\langle A\rangle = \langle\psi(t)| \hat{A} |\psi(t)\rangle . \tag{3.3.6}$$

The wavefunction $\psi(t)$ is assumed to be normalized and well defined.

Robertson subsequently generalized this result for a pair of self-adjoint operators $\hat{A}$ and $\hat{B}$:

$$\sigma(A)\sigma(B) \geq \frac{1}{2} \left|\langle\psi(t)| \left(\hat{A}\hat{B} - \hat{B}\hat{A}\right) |\psi(t)\rangle\right| . \tag{3.3.7}$$

Substituting $\hat{A} = \hat{q}$ and $\hat{B} = \hat{p} = \frac{\hbar}{i}\partial_q$ yields Eq. (3.3.4). Since $\sigma(Q)$ and $\sigma(P)$ are standard deviations of measurable statistical quantities, they are, in principle, experimentally accessible. Robertson's formulation clarifies that the mathematical origin of the uncertainty relation lies in the noncommutativity of operators.

Schrödinger later refined these relations, obtaining

$$\begin{aligned} \sigma(A)^2\sigma(B)^2 \geq &\left|\frac{1}{2} \langle\psi(t)| \left(\hat{A}\hat{B} - \hat{B}\hat{A}\right) |\psi(t)\rangle\right|^2 \\ &+ \left(\frac{1}{2} \langle\psi(t)| \left(\hat{A}\hat{B} + \hat{B}\hat{A}\right) |\psi(t)\rangle - \langle A\rangle \langle B\rangle\right)^2 . \end{aligned} \tag{3.3.8}$$

These are collectively referred to as the KRS uncertainty relations. (Even in cases where self-adjoint operators do not exist, uncertainty relations may still hold, as in the case of energy and time; see Chap. 9.)

Notably, the KRS uncertainty relations do not incorporate the disturbance induced by measurement and are therefore distinct from Heisenberg's relation Eq. (3.3.3). Conversely, Heisenberg's relation does not explicitly include statistical dispersion. Ozawa unified these viewpoints within the framework of quantum measurement theory and derived the generalized uncertainty relation as

$$\varepsilon(Q)\eta(P) + \varepsilon(Q)\sigma(P) + \sigma(Q)\eta(P) \geq \frac{\hbar}{2}, \tag{3.3.9}$$

known as Ozawa's inequality [52, 53]. A notable feature of this relation is that, depending on the measurement scheme, Heisenberg's original relation Eq. (3.3.3) may be violated, implying that the measurement precision $\varepsilon(Q)$ is not fundamentally constrained by it exclusively.

Thus, it has been demonstrated that measurement processes introduce irreversible disturbances into quantum systems, while independently, uncertainty relations also emerge from the mathematical properties of wavefunctions and operators. Furthermore, as noted above, uncertainty relations are not exclusive to quantum theory. This raises the question of the possible physical origin underlying the KRS uncertainty relations, which will be discussed in Chap. 9.

3.3.5 Nonlocal Correlations and Quantum Entanglement

3.3.5.1 Overview of the Entanglement

Quantum entanglement (simply referred to as entanglement) is a familiar and routinely encountered concept for researchers dealing with many-electron systems in atomic, molecular, and condensed matter physics. It also plays an important role in processes such as the intertwining of photons. However, in this book, we focus specifically on particle systems, such as electrons, atomic nuclei, and those composed thereof, and treat them separately from photonic systems. Wavefunctions in such cases that can only be expressed in a sum-of-products form, such as

$$\psi(\vec{Q}_1, \vec{Q}_2, t) = \sum_i c_i(t)\chi_i(\vec{Q}_1)\xi_i(\vec{Q}_2) \tag{3.3.10}$$

may appear unusual, yet they are legitimate solutions of the Schrödinger equation under the superposition principle. As time elapses, such forms may factorize or may not converge to a simple product form, often due to symmetry requirement. This defines an entangled state.

For example, since electrons are fermions, their wavefunctions must be antisymmetric with respect to permutation (exchange). In a two-electron system, for instance, the spin functions are restricted to the singlet:

$$\Theta_1(1, 2) = \alpha(1)\beta(2) - \beta(1)\alpha(2) \tag{3.3.11}$$

and the triplet

$$\Theta_3(1, 2) = \alpha(1)\beta(2) + \beta(1)\alpha(2). \tag{3.3.12}$$

The former is antisymmetric and the latter symmetric under permutation. These forms cannot be reduced to simple products due to the symmetry of the Schrödinger equation itself. The manner of entanglement significantly influences observable physical properties: For instance, the ground state of the helium atom is a singlet, whereas triplet states lie at much higher energies. Similarly, the ground state of the oxygen molecule we breathe is a triplet, exhibiting properties markedly different from the corresponding singlet state. Thus, modern materials science is fundamentally dependent on entanglement.

As is well known from the fact that his name is attached to Bose-Einstein statistics, Einstein was thoroughly familiar with the permutation symmetry of particles and its physical consequences. Therefore, he would have had no objection to the essential role played in quantum theory by the permutation symmetry of fermions—for example, the distinction between $\Theta_1(1, 2)$ and $\Theta_3(1, 2)$. However, Einstein and his collaborators [54] regarded as a "defect of quantum mechanics" the implication that, once a pair of electrons has formed a singlet state, the two electrons must remain "correlated" so long as they obey the Schrödinger equation, even if they are separated by a large distance. For instance, suppose that a singlet pair of electrons is created in Honolulu and subsequently separates such that one electron is observed in Tokyo and the other in Los Angeles. According to the Schrödinger equation, the two electrons remain in a singlet state. In this situation, the spin of each electron to be detected—whether α-spin or β-spin—has a probability of 1/2 in both Tokyo and Los Angeles. Now, assume that a measurement performed only in Tokyo yields α-spin. Then, it necessarily follows that the electron in Los Angeles must be in the β-spin state. However, from the perspective of Los Angeles (LA), both α(Tokyo)β(LA) and β(Tokyo)α(LA) should be equally probable; nevertheless, only α(Tokyo)β(LA) is realized, effectively selected from the two equally likely possibilities without any measurement being performed in Los Angeles. Thus, even though no action has been taken in Los Angeles, the spin state there is determined, implying that the observers in Tokyo and Los Angeles have, without awareness or intention, become correlated through the specific relation α(Tokyo)β(LA).

Of course, since this situation becomes apparent only after a measurement is made, some have even described it by saying that "the act of measurement itself selects one outcome," thereby interpreting it as a kind of wavefunction collapse. The use of Tokyo and Los Angeles here is merely illustrative; the point is to emphasize that, irrespective of the distance between observers, the two systems become related in a definite manner without conscious intervention. Moreover, if one imagines that such information might have been transmitted faster than the speed of light, it becomes evident that the issue is not merely one of spatial separation. (Incidentally, it has already been proved that entanglement cannot be used as a means to transfer information faster than the speed of light.)

The fact that a measurement performed in Tokyo correlates, in some manner, with the situation in Los Angeles exemplifies what is called quantum non-locality—a connection that persists irrespective of distance and manifests as a kind of background linkage. (Conversely, the principle that observations or phenomena in spatially separated systems cannot influence each other faster than the speed of light is referred to as locality.) A collection of particles in an entangled state may thus be regarded as constituting an inseparable single system, regardless of the distance between them, irrespective of the presence of Coulomb interactions and so on. They remain in a state of correlation that cannot be understood in terms of conventional notions of proximity interactions. Such behavior is, of course, inconceivable within classical mechanics, which is founded on the principle of locality. Indeed, it is not even clear whether these long-range correlations arise from any physical interaction in the usual sense. Einstein referred to this as a "spooky" interaction and claimed based on the following mathematical argument that quantum theory must either be incomplete or contain some errors.

The concrete example pointed out by Einstein, Podolsky, and Rosen in their so-called EPR paper, as a consequence of the persistence of entanglement even at arbitrarily large separations, is an apparent conflict with the uncertainty principle [54]. To briefly outline the argument, consider two electrons prepared in an entangled state such that the position of their center of mass, $(q_1 + q_2)/2$, and the difference of their momenta, $(p_1 - p_2)$, are both known. The operators $(\hat{q}_1 + \hat{q}_2)$ and $\left(\hat{p}_1 - \hat{p}_2\right)$ formally commute and hence are not subject to an uncertainty relation. When the two electrons subsequently move in opposite directions, one may measure the position of one particle (denoted q_1^{∞}) and the momentum of the other (denoted p_2^{∞}) with arbitrary precision. By straightforward application of conservation laws, one can then infer q_2^{∞} and p_1^{∞} at the same instant. In this way, both pairs $(q_1^{\infty}, p_1^{\infty})$ and $(q_2^{\infty}, p_2^{\infty})$ can be determined precisely, suggesting that the uncertainty relation—one of the central principles of quantum theory—would not hold. In other words, this constitutes a mathematical argument that the uncertainty relation could be violated by exploiting entanglement.

On the basis of the notion of "physical reality," Einstein and his collaborators raised doubts about the uncertainty relation, which asserts that observable quantities such as position and momentum cannot simultaneously possess definite values (or we cannot simultaneously give the individual value to each). They argued that, since entanglement—arising as a consequence of the superposition principle of the wave function—appears to enable a violation of the uncertainty relation, quantum mechanics must be either incomplete or in some respect incorrect. From this perspective, entanglement and the uncertainty relation, both derived within the framework of the Schrödinger equation, do not seem to be mutually compatible.

Incidentally, by physical reality, EPR paper means "If, without in any way disturbing a system, we can predict with certainty (i.e., with probability equal to unity) the value of a physical quantity, then there exists an element of physical reality corresponding to this physical quantity." Mathematically, both entanglement and the uncertainty relation are well founded within the formalism of the Schrödinger equation. Besides, the validity of the uncertainty relation has been proved experimentally. Therefore, if one regards entanglement as having been empirically established—particularly through the violation of Bell's inequality discussed below—one encounters profound conceptual difficulties. Although the notion of "physical reality" constitutes an important premise, it carries strong philosophical and rhetorical implications and will not be pursued in this book. We hence skip this aspect but rather get involved in the mechanism of uncertainty relations and the spontaneous disentanglement inherent to the quantum stochastic path dynamics in Chap. 9.

Here, it would be appropriate to add a supplementary remark concerning the EPR paper. The wave function they presented as an example that could, in principle, violate the uncertainty relation is of the form:

$$\Psi_{EPR}(q_1, q_2) = \int_{-\infty}^{\infty} \exp\left[\frac{i}{\hbar}(q_1 - q_2 + q_0)p\right] dp, \tag{3.3.13}$$

where q_0 plays the role of the eigenvalue of the operator $\hat{q}_2 - \hat{q}_1$. In the EPR paper, it is argued that by measuring q_1 and p_1, one can determine $\left(q_2^{\infty}, p_2^{\infty}\right)$ without any interaction with the second particle. (Note that since there is no time coordinate in this argument, no restriction on simultaneity seems to be imposed.) Although such a mathematical procedure can indeed be formally carried out, the author finds it difficult to envisage how one should realize the corresponding experimental initial conditions for $\Psi_{EPR}(q_1, q_2)$. Besides, in general discussions of entanglement, it is customary that neither the time evolution nor the time dependence of the system is explicitly addressed. Indeed, the function $\Psi_{EPR}(q_1, q_2)$ itself contains no time variable. However, the author thinks that entanglement can/may transform to spontaneous disentanglement—namely, an intrinsic time dependence that provides a dynamical "exit (and entrance as well)." For this reason, this issue will be examined in detail in Chap. 9 within the framework of individual quantum path dynamics.

In quantum theory as applied to materials science, one does not usually consider entanglement in systems of planetary or cosmological scale, and Einstein's criticism might appear somewhat overstated. However, abandoning the principle of locality, which claims that any physical interaction cannot transmit faster than the speed of light, would undermine the very possibility of isolating and analyzing physical systems, potentially calling into question the nature of physical laws themselves. Furthermore, if such "spooky" interactions were to persist across the entire universe, they would have to be incorporated into the entire framework of causality, leading to the profound issue of non-separability. This is a problem of both vast scope and deep significance—one that Einstein could hardly have accepted.

3.3.5.2 Experimental Verification of Non-locality: Bell's Inequality

The distinction between locality and non-locality can be experimentally tested, as proposed by Bell's inequality in 1964 [41]. This inequality has been modified in various ways. Since there have been many explanatory books and articles published, please refer to those for details. Here, we will only briefly summarize the essential content common to the various forms of Bell's inequality as follows.

The explanation organized concisely in "The Worldview of Quantum Mechanics" (edited by Hiroshi Ezawa and Takashi Kamijo, Nihon Hyoronsha, 2019) is based directly on the original paper [41] and that of Clauser, Horne, Shimony, and Holt [55] and succinctly summarizes the essentials. Here, we follow it.

Suppose a performance of measurement a and the value $A(a)$ is obtained. Similarly, an independent measurement b yields the value $B(b)$. Next, assume that there is some interaction between these two measurements, which causes a correlation between them. The magnitude and type of this interaction are represented by a parameter λ, and the probability of its occurrence is denoted by a function $\rho(\lambda)$, which is required that $\rho(\lambda) \geq 0$ and $\int \rho(\lambda)d\lambda = 1$. This interaction acts directly (that is, locally) on the measurement systems a and b. Furthermore, from the standpoint of theoretical assumptions, the interaction represented by the parameter λ is not limited to known phenomena. It may also include phenomena that are not yet known (hidden from human knowledge).

Under these conditions, the two measurements are not independent, but are both influenced by some common interaction, so their measurement values become functions of λ, namely, $A(a, \lambda)$ and $B(b, \lambda)$. Furthermore, due to the existence of λ, statistical correlations arise between the two. The corresponding relationship is to be written as

$$P(a, b) = \int A(a, \lambda)B(b, \lambda)\rho(\lambda)d\lambda. \tag{3.3.14}$$

Here, we may assume without losing generality that

$$|A(a, \lambda)| \leq 1, \qquad |B(b, \lambda)| \leq 1. \tag{3.3.15}$$

Then, Bell showed [41] that it naturally follows that the quantity defined as

$$S = P(a, b) - P(a, b') + P(a', b) + P(a', b') \tag{3.3.16}$$

should always satisfy

$$|S| \leq 2. \tag{3.3.17}$$

If this inequality is violated, the classical physical depiction that "the two correlated systems are each influenced only by some local action" is broken.

The fact that Bell's inequality is violated in quantum systems was demonstrated by the ingenious experiments of Clauser, Aspect, and Zeilinger. The experiments they conducted are believed to have evidenced:

1. In response to the hypothesis that quantum theory can be rendered complete by supplementing it with hidden variables, it has been established that no local hidden variables exist.
2. Nonlocal correlations can exist.

In fact, it can be proven using only the theoretical framework of quantum mechanics that Bell's inequality can be violated within quantum theory. Explanations of Bell's inequality and its violation, including historical aspects, are given in detail in the book by Home and Whitaker [16]. Since proofs of the violation are presented thoroughly in works such as Akira Shimizu's *Foundations of Quantum Theory* (New Edition, Science-sha), we do not reproduce them here; however, many (analytical) proofs make use of the relative angle between two electron spins.

Implicit in such arguments is the assumption that entanglement does not depend on the spatial separation between particles, but rather arises from the mathematical structure (symmetry) of the wavefunction. Accordingly, the proof is carried out independently of the distance between the electrons, leading to the conclusion that correlations operative at atomic and molecular scales persist over arbitrarily large distances, regardless of their magnitude. Thus, the question of what actually occurs when electrons are separated by large distances is not itself made an object of discussion.

As noted above, without accepting entanglement, the field of materials science could not be sustained. The physical phenomenon of entanglement undoubtedly exists. However, is the peculiar long-range correlation that concerned Einstein truly of the same nature as the entanglement observed at atomic and molecular scales?

3.3.5.3 Did Einstein Accept It?

Einstein passed away in 1955 and therefore had no opportunity to know of Bell's inequality (1964) or the experiments carried out by John Clauser in 1972. Thus, the question of whether Einstein would have been convinced by the demonstration of non-locality or non-separability in quantum theory is inherently hypothetical.

To restate Einstein's concern: Once a system enters an entangled state, even if its components become geometrically separated by large distances, a peculiar, non-separable correlation persists, potentially allowing the domain of causal relations to expand without bound. Even if Bell's inequality is violated, it is unlikely that Einstein would have accepted a line of reasoning that undermines the fundamental framework of science.

3.3.5.4 Chemical Bond and Entanglement

Let us consider another very simple example. Suppose that the hydrogen molecular ion H_2^+, composed of one electron and two protons, exists and, for instance, dissociates into a hydrogen atom and a proton through a physical method like photoexcitation. Assume that this dissociation process occurs in a completely symmetric manner, giving rise to the following two channels (possible outcomes):

$$(\mathrm{H_A - H_B})^+ \rightarrow \begin{array}{l} \nearrow \mathrm{H_A^+ + H_B} \\ \searrow \mathrm{H_A + H_B^+} \end{array} \tag{3.3.18}$$

and suppose that these occur with exactly equal probability. However, if the dissociation is carried out while maintaining symmetry in accordance with the requirements of the Schrödinger equation, neither of the individual channels in Eq. (3.3.18) constitutes a solution by itself; rather, their superposition forms the solution. In this sense, it is not entirely incorrect to say that $H_A^{0.5+} + H_B^{0.5+}$ represents the electron density obtained from the Schrödinger equation. In a shorthand formula, one sometimes writes this situation as

$$(H_A - H_B)^+ \rightarrow H_A^{0.5+} + H_B^{0.5+}. \tag{3.3.19}$$

(For convenience, the atoms have been labeled, but exchange symmetry is assumed to be preserved.) Of course, there is no physically realizable channel in which the electron is literally split in half to produce $H_A^{0.5+} + H_B^{0.5+}$. This is because the entanglement between the electron and the nuclei is maintained even in the dissociation limit. While this charge distribution is correct as an average description, in each individual dissociation event, the appearance of entities such as $H^{0.5+}$ is clearly unphysical and cannot be observed. How do you teach this counterintuitive "phenomenon" in the introductory courses on chemical bonding?

3.3.5.5 Exit and Entrance

Considering the above example, the symmetry assumed in the Schrödinger equation—such as that expressed in Eq. (3.3.19)—must, in actual physical processes like those described by Eq. (3.3.18), be broken at some stage. This observation strongly suggests that there is an inherent ambiguity in the commonly stated claim that "once a system enters an entangled state, a peculiar, non-separable correlation persists even when the subsystems are geometrically far apart, and the domain of causal relations continues to expand without bound."

Therefore, we may ask when, where, and how could a system that was initially separable enter an entangled state? From the time-reversal symmetry of the Schrödinger equation, one would expect that if a system can enter an entangled state, there must also exist channels through which it can exit such a state. If there is an entrance to entanglement, there should also be an exit. (Note: We are not considering the entanglement of photons here.) If neither an entrance nor an exit exists, then the universe would have remained in a tightly entangled state since its origin. In that case, it would be meaningless to isolate an extremely small subsystem (for example, the notion of a single hydrogen atom) from the universe and consider the Schrödinger equation for it. Rather, one would be led to require a theory such as a "wavefunction of the entire universe."

Note that Bell's theorem itself is not a theory of quantum mechanics. Although it is undoubtedly a profound idea, its logic is restricted to general statistical properties—namely, that two phenomena governed (whether explicitly or implicitly) by underlying local interactions must satisfy certain conditions when observed. As such, it does not, by itself, provide a means of answering the fundamental questions of quantum mechanics. In this book, we shall examine entanglement and its spontaneous breaking from the standpoint of quantum theory in Chap. 9.

3.3.6 Categories of Interpretations of the Schrödinger Equation

Up to this point, as part of the motivation for the new theoretical developments beginning in the next chapter, we have introduced only the most basic issues in quantum theory and a selection of interpretational approaches intended to address them. (The author braces himself for criticism from specialists in the field of "interpretations.") Interpretations of quantum theory span a wide spectrum, ranging from philosophical considerations to purely mathematical frameworks such as quantum logic.

Here, based on the examples discussed in this chapter, the author proposes to classify interpretations related to Schrödinger dynamics that are open to question into the following three categories. This classification is not intended as a critique of past interpretations or hypotheses, but rather as groundwork for advancing the present theory of Schrödinger dynamics.

1. Interpretations based on unsupported assumptions about the wavefunction (i.e., assumptions that cannot be established either theoretically or experimentally)

   Examples
   Sudden collapse of the wavefunction
   The interpretation in the double-slit experiment that a particle can pass through two slits simultaneously (or has the possibility of doing so) and interferes with itself
   The wavefunction of the universe and the branching of worlds

2. Arguments relying solely on the wavefunction (i.e., claims about phenomena or properties that cannot, in principle, be fully described by the wavefunction alone)

   Examples
   The existence of peculiar long-range correlations associated with entanglement
   Schrödinger's cat (proposed as a supplement to the EPR paper [54] to highlight the peculiarities of Schrödinger dynamics)

3. Incomplete understanding of quantum phenomena arising from the absence of a quantum theory for individual events

   Examples
   Sudden collapse of the wavefunction
   The notion that a particle simultaneously passes through two slits and produces self-interference
   Random arrival of each particle in the double-slit experiment

3.4 Looking Ahead for the Better Understanding

In concluding this chapter, we seek directions for reconstructing a new fundamental framework for the theory of Schrödinger dynamics.

3.4.1 Is the Wavefunction Everything?

As explicitly noted in the discussion of the double-slit experiment in Chap. 2, the Schrödinger wavefunction represents, by a single function, a multitude of events that may arise over many repeated trials. Conversely, it is insufficient for describing or explaining individual events.

3.4.1.1 Risk of Relying Solely on the Schrödinger Equation and the Wavefunctions

Historically, debates concerning the interpretation of quantum theory—such as the Einstein-Bohr controversy—appear to have been conducted under the implicit restriction that quantum dynamics is equivalent to the Schrödinger equation. This restriction was perhaps unavoidable at the time. However, attempting to understand and explain all quantum phenomena solely in terms of the Schrödinger equation and the wavefunction is, in essence, an overreach. Even if the Schrödinger equation is mathematically self-consistent and closed, it does not necessarily follow that all quantum phenomena are fully captured by the wavefunction, nor does there appear to be any means of proving such a claim. In particular, invoking the wavefunction of the universe risks falling into an unprovable monistic view of the cosmos.

3.4.1.2 Path Representation to Express Quantum Dynamics

With the transition from the Newtonian equation to the Schrödinger equation as the fundamental equation of the microscopic world, the description shifted from trajectories to wavefunctions. In this context, Albert Einstein is said to have asked Werner Heisenberg: "In a Wilson cloud chamber, one can clearly trace the trajectory of an electron; why, then, is the concept of an orbit denied within an atom?" Heisenberg reportedly replied: "Within an atom, electron orbits cannot be observed. ⋯ A theory should concern only what can, in principle, be observed." (W. Heisenberg, *Der Teil und das Ganze*). From the present perspective, this response appears not to directly address Einstein's question or, at best, to shift the focus. Moreover, the idea that "a theory should concern only observable quantities" was subsequently sharpened into the view that not only does observation reveal the world, but also that only what is observed possesses physical meaning.

With regard to the path concept, Richard Feynman proposed in his 1948 paper [56] that the Green's function of the Schrödinger equation can be expressed as a sum over all conceivable paths. Each path is associated with a complex amplitude of magnitude corresponding to the quantum phase derived from the action integral, and the superposition of contributions from infinitely many such paths allows the time evolution of the wavefunction to be visualized not as a "cloud" but as an ensemble of "all the possible path lines." This formulation is now widely known as the path integral method. However, since it lacks a proper measure as an integral, the absolute convergence is not guaranteed; this issue will be discussed in detail in Chap. 6. The connection between the action integral, the Lagrangian formalism, and both quantum and classical mechanics was already suggested in the textbook of Dirac [1], which Feynman himself acknowledged as guidance toward the development of the path integral.

Related aspects were also discussed in connection with the Hamilton-Jacobi equation (3.2.9). In 1952, David Bohm, starting from his wavefunction representation (Eq. (3.2.20)), introduced what are now known as Bohmian trajectories. Although, like the Feynman formulation, they are guided by the action integral, Bohm sought to assign physical meaning to each individual trajectory governed by dynamical laws incorporating quantum effects.

Going further, one may ask whether there exists an equation of motion capable of "unfolding" the individual events contained within the wavefunction and predicting their respective dynamics. This problem was investigated in mathematical detail by Masao Nagasawa [57, 58], whose formulation takes the form of stochastic differential equations for quantum paths. In this book as well, we propose, in Chap. 8, a theoretical framework for the quantum dynamics of individual events.

3.4.2 Focus on the Reconstruction of Quantum Dynamics

In light of the issues outlined above concerning quantum theory and its fundamental interpretations, we itemize again the problems listed in Chap. 2 in preparation for the challenges that follow.

3.4.2.1 Re-examining the Nature of the Wavefunction

- The meaning and role of the Schrödinger equation.
- What is the wavefunction, fundamentally?
- The origin of the Born probability interpretation.
- Does the wavefunction collapse?
- Does the wavefunction represent a matter wave?
- Is the de Broglie matter wave an intrinsic, physically real property of matter?
- What is meant by de Broglie's wave-particle duality?

3.4.2.2 Possibility of Individual Path Dynamics

- Consistency between statistical summation and the quantum superposition principle.
- Why do individually emitted electrons form interference patterns in the double-slit experiment?
- The relationship between quantum dynamics and classical dynamics (Hamilton's canonical equations, Newtonian equations).
- The kinematical origin of the uncertainty relation.
- Preservation and breaking of symmetries in the Schrödinger equation.

- Quantum entanglement as a "spooky" long-range correlation and the possibility of spontaneous disentanglement.

References

1. Dirac, P. A. M. *The Principles of Quantum Mechanics* (Oxford university press, 1981).
2. Von Neumann, J. *Mathematical Foundations of Quantum Mechanics: New edition*, vol. 53 (Princeton university press, 2018).
3. Schiff, L. I. *Quantum Mechanics* (McGraw–Hill, New York, 1968).
4. Aharonov, Y., Bergmann, P. G. & Lebowitz, J. L. Time symmetry in the quantum process of measurement. *Physical Review* **134**, B1410 (1964).
5. Reznik, B. & Aharonov, Y. Time-symmetric formulation of quantum mechanics. *Physical Review A* **52**, 2538 (1995).
6. Aharonov, Y., Popescu, S. & Tollaksen, J. A time-symmetric formulation of quantum mechanics. *Physics Today* **63**, 27–32 (2010).
7. Messiah, A. *Quantum Mechanics* (Courier Corporation, 2014).
8. Landau, L. D. & Lifshitz, E. M. *Quantum Mechanics: Non-relativistic Theory*, vol. 3 (Elsevier, 2013).
9. Cohen-Tannoudji, C., Diu, B. & Laloë, F. *Quantum Mechanics* (John Wiley & Sons, 2019).
10. Weinberg, S. *Lectures on Quantum Mechanics* (Cambridge University Press, 2015).
11. Sakurai, J. J. & Napolitano, J. *Modern Quantum Mechanics* (Cambridge University Press, 2020).
12. Feynman, R. P., Hibbs, A. R. & Styer, D. F. *Quantum Mechanics and Path Integrals* (Courier Corporation, 2010).
13. Wigner, E. *Group Theory: And its Application to the Quantum Mechanics of Atomic Spectra*, vol. 5 (Elsevier, 2012).
14. Newton, R. G. *Scattering Theory of Waves and Particles* (Springer Science & Business Media, 2013).
15. David Mermin, N. What's wrong with this pillow? *Physics Today* **42**, 9–11 (1989).
16. Home, D. & Whitaker, A. *Einstein's Struggles with Quantum Theory: A Reappraisal* (Springer Science & Business Media, 2007).
17. Kumar, M. *Quantum: Einstein, Bohr and the Great Debate about the Nature of Reality* (Icon Books Ltd, 2008).
18. Nelson, E. Derivation of the schrödinger equation from newtonian mechanics. *Physical Review* **150**, 1079 (1966).
19. Arnold, V. I. *Mathematical Methods of Classical Mechanics* (Springer, Berlin, 1989), 2 edn.
20. Bohm, D. A suggested interpretation of the quantum theory in terms of "hidden" variables. I. *Physical Review* **85**, 166 (1952).
21. Maslov, V. P. & Fedoriuk, M. V. *Semi-Classical Approximation in Quantum Mechanics*, vol. 7 (Springer Science & Business Media, 2001).
22. Takatsuka, K. & Inoue, A. Global representation of Maslov-type semiclassical wave function and its spectrum in a small number of classical trajectories. *Physical Review Letters* **78**, 1404 (1997).
23. Inoue-Ushiyama, A. & Takatsuka, K. Semiclassical theory for the Maslov-type wave packet: Hierarchy below the semiclassical Feynman kernel. *Physical Review A* **59**, 3256 (1999).
24. Takatsuka, K. Amplitude-free correlation function based on an algebra for coordinate transformation in semiclassical integrals. *Physical Review E* **64**, 016224 (2001).
25. Takahashi, S. & Takatsuka, K. Towards many-dimensional real-time quantum theory for heavy-particle dynamics. I. semiclassics in the lagrange picture of classical phase flow. *Physical Review A* **89**, 012108 (2014).

26. Takatsuka, K. & Takahashi, S. Towards many-dimensional real-time quantum theory for heavy-particle dynamics. ii. beyond semiclassics by quantum smoothing of the singularity in quantum-classical correspondence. *Physical Review A* **89**, 012109 (2014).
27. Wigner, E. On the quantum correction for thermodynamic equilibrium. *Physical Review* **40**, 749 (1932).
28. Louisell, W. H. *Quantum Statistical Properties of Radiation* (John Wiley and Sons, Inc., New York, 1973).
29. Lee, H.-W. Theory and application of the quantum phase-space distribution functions. *Physics Reports* **259**, 147–211 (1995).
30. Zachos, C., Fairlie, D. & Curtright, T. *Quantum Mechanics in Phase Space: an overview with selected papers* (World Scientific, 2005).
31. Carruthers, P. & Zachariasen, F. Quantum collision theory with phase-space distributions. *Reviews of Modern Physics* **55**, 245 (1983).
32. Polkovnikov, A. Phase space representation of quantum dynamics. *Annals of Physics* **325**, 1790–1852 (2010).
33. Takatsuka, K. Phase-space path integrals in terms of a phase-space distribution function. *Physical Review Letters* **61**, 503 (1988).
34. Takatsuka, K. Phase-space representation of quantum mechanics and its relation to phase-space path integrals. *Physical Review A* **39**, 5961 (1989).
35. Boesten, L., Tanaka, H., Kobayashi, A., Dillon, M. & Kimura, M. Crossed-beam experiment for the scattering of low energy electrons from cf4. *Journal of Physics B: Atomic, Molecular and Optical Physics* **25**, 1607 (1992).
36. Isaacs, W., McCurdy, C. & Rescigno, T. Theoretical support for a Ramsauer-Townsend minimum in electron-cf 4 scattering. *Physical Review A* **58**, 309 (1998).
37. Gianturco, F. & Willner, K. Ramsauer-townsend effect for electron scattering from gaseous cf 4 molecules. *Physical Review A Atomic, Molecular, and Optical Physics* **75**, 062714 (2007).
38. Omnes, R. Consistent interpretations of quantum mechanics. *Reviews of Modern Physics* **64**, 339 (1992).
39. Auletta, G. *Foundations and Interpretation of Quantum Mechanics: In the Light of a Critical-Historical Analysis of the Problems and of a Synthesis of the Results* (World Scientific, 2001).
40. Freire Jr, O. *et al. The Oxford Handbook of the History of Quantum Interpretations* (Oxford University Press, 2022).
41. Bell, J. S. On the Einstein Podolsky Rosen paradox. *Physics Physique Fizika* **1**, 195 (1964).
42. Kochen, S. & Specker, E. P. The problem of hidden variables in quantum mechanics. *Journal of Mathematics and Mechanics* **17**, 59–87 (1967).
43. t Hooft, G. *The Cellular Automaton Interpretation of Quantum Mechanics* (Springer Nature, 2016).
44. Ghirardi, G. C., Rimini, A. & Weber, T. Unified dynamics for microscopic and macroscopic systems. *Physical Review D* **34**, 470 (1986).
45. Bassi, A., Lochan, K., Satin, S., Singh, T. P. & Ulbricht, H. Models of wave-function collapse, underlying theories, and experimental tests. *Reviews of Modern Physics* **85**, 471–527 (2013).
46. Everett III, H. "Relative state" formulation of quantum mechanics. *Reviews of Modern Physics* **29**, 454 (1957).
47. DeWitt, B. S. Quantum mechanics and reality. *Physics Today* **23**, 30–35 (1970).
48. Camilleri, K. A history of entanglement: Decoherence and the interpretation problem. *Studies in History and Philosophy of Science Part B: Studies in History and Philosophy of Modern Physics* **40**, 290–302 (2009).
49. Schlosshauer, M. Decoherence, the measurement problem, and interpretations of quantum mechanics. *Reviews of Modern Physics* **76**, 1267–1305 (2004).
50. Schlosshauer, M. Quantum decoherence. *Physics Reports* **831**, 1–57 (2019).
51. Mermin, N. D. Is the moon there when nobody looks? reality and the quantum theory. *Physics Today* **38**, 38–47 (1985).
52. Ozawa, M. Universally valid reformulation of the Heisenberg uncertainty principle on noise and disturbance in measurement. *Physical Review A* **67**, 042105 (2003).

53. Ozawa, M. Heisenberg's original derivation of the uncertainty principle and its universally valid reformulations. *Current Science* 2006–2016 (2015).
54. Einstein, A., Podolsky, B. & Rosen, N. Can quantum-mechanical description of physical reality be considered complete? *Physical Review* **47**, 777 (1935).
55. Clauser, J. F., Horne, M. A., Shimony, A. & Holt, R. A. Proposed experiment to test local hidden-variable theories. *Physical Review Letters* **23**, 880 (1969).
56. Feynman, R. P. Space-time approach to non-relativistic quantum mechanics. *Reviews of Modern Physics* **20**, 367 (1948).
57. Nagasawa, M. Transformations of diffusion and schrödinger processes. *Probability Theory and Related Fields* **82**, 109–136 (1989).
58. Nagasawa, M. *Schrödinger Equations and Diffusion Theory* (Springer Science & Business Media, 2012).

Stochastic Processes: Diffusion Equation, Feynman-Kac Formula, and Ito Stochastic Differential Equation

4

Abstract

This book claims that the Schrödinger dynamics has a three-pillared structure as a fundamental structure and that each of three pillars has its counterpart in stochastic (diffusion) dynamics. Those are (1) diffusion equation, (2) Wiener path integral or the Feynman-Kac formula, and (3) Ito's stochastic differential equation. In order to understand the similarity and difference between the Schrödinger dynamics and stochastic dynamics, we first present a concise review of diffusion dynamics. In particular, a short take a short look at the relationship between the parabolic partial differential equation and the stochastic process is given.

Keywords

Diffusion equation · Wigner path integral · Feynman-Kac formula · Ito stochastic differential equation · Velocity field · Parabolic partial differential equation

After stating the objective of this book in Chap. 2, we discussed in Chap. 3 the underlying motivations and the necessity for new developments. We now turn to the main subject, and in this Chap. 4, we prepare the theoretical groundwork required for what follows.

The reader is invited to revisit Fig. 2.2 presented in Chap. 2. In this book, Schrödinger dynamics is formulated in terms of three mutually correlated theoretical pillars, for each of which new governing equations are derived. Based on this framework, we aim to address the fundamental issues of quantum theory discussed in the preceding chapters. As indicated in Fig. 2.2, these three pillars of Schrödinger dynamics correspond directly to three pillars in the theory of stochastic processes. One of the central claims of this book is that quantum theory may be regarded

K. Takatsuka, *Structure of the Schrödinger Dynamics*,
https://doi.org/10.1007/978-981-92-1910-0_4

as describing dynamics at a different hierarchical level than stochastic process theory. Accordingly, as a pathway into quantum theory, we begin by reviewing the fundamentals of stochastic processes.

As will be developed sequentially from Chap. 5, the mathematical formulations corresponding to the three elements of quantum dynamics shown in Fig. 2.2, together with their derivations, are all based on the results reported in Ref. [1]. The theory to be presented in detail in Chap. 8 is also due to the author. However, as far as the profound relationship between quantum theory and stochastic differential equations, which in turn is rooted in the correspondence between parabolic partial differential equations and stochastic processes, is concerned, Masao Nagasawa pioneered this area, whose English monograph *Schrödinger Equations and Diffusion Theory* provides a comprehensive account [2]. In Japan, his work is also accessible through *A New Quantum Theory Based on Markov Processes* (2012). Furthermore, developments connecting stochastic processes to quantum theory are presented in *Brownian Motion* by Hiroshi Ezawa and Toru Nakamura [3], both of which are authoritative and insightful contributions.

In this chapter, we explain only the basic aspects of stochastic processes and statistical mechanics necessary for the present work. Given that Japan has produced many outstanding contributions to stochastic process theory—nurtured in part by the work of Kiyosi Ito the originator of stochastic differential equations—the reader is encouraged to consult not only the two references mentioned above but also other related literature.

4.1 Langevin Dynamics: Abandoning the Phase Space View

As is well known, the Langevin equation describing the Brownian motion of pollen particles suspended on the surface of water takes the form:

$$\frac{dp}{dt} = -\frac{\partial V(q)}{\partial q} - \gamma p + R(t), \tag{4.1.1}$$

where p denotes the momentum of the particle, $-\gamma p$ represents the frictional force, and $R(t)$ is a random force exerted by the surrounding water molecules. The essential differences from Newtonian mechanics are

1. The abandonment of the phase space $(q - p)$ description of dynamics
2. The explicit introduction of stochasticity into the dynamics via the random force $R(t)$

The Langevin equation was proposed in 1908, following Einstein's theory of Brownian motion (1905), well before the completion of quantum theory by Schrödinger and others. These two features would later become directly relevant to quantum theory.

The mathematical structure underlying the Langevin equation was subsequently formalized, leading to the development of stochastic differential equations by Ito and others in the early 1940s. Notably, the emergence of the Schrödinger equation (1925–1926) lies chronologically between Langevin's work and Ito's formulation. As a consequence, early developments in quantum theory—including the Copenhagen interpretation—proceeded without recognizing the relationship between partial differential equations and stochastic processes [2].

4.2 Diffusion Equation: Conservation of Distribution and Relevant Flux

By tracing the trajectory of a Brownian particle, we express the probability that it reaches a certain point q after time t by the distribution function $\rho_d(q, t)$. We have abandoned a phase space ($q - p$ space) description, and what is observed under a microscope is the position of each sample q rather than the associated momentum p. Therefore, unlike the illustration of the Langevin equation (4.1.1), it is natural to represent the distribution function as $\rho_d(q, t)$ in terms of q. From the view point of the Liouville function $\Gamma(q, p, t)$, Eq. (3.2.34), the p-space variables may be regarded as integrated out. However, just as with $\Gamma(q, p, t)$, the number of Brownian particles is necessarily conserved in $\rho_d(q, t)$, so the pure diffusion process forms a noncompressible flow in q-space. Therefore, due to the Gauss theorem, there exists a flux $\vec{j}$ that should satisfy the equation of continuity:

$$\frac{\partial}{\partial t}\rho_d(q, t) = -\vec{\nabla} \cdot \vec{j}. \tag{4.2.1}$$

Here, the intuitive (entropic) idea that the flux flows from regions of high density to regions of low density suggests that it is proportional to the gradient of the density. If we denote the proportionality constant at this time as D, then we have

$$\vec{j} = -D\vec{\nabla}\rho_d \tag{4.2.2}$$

and from Eq. (4.2.1), it holds that

$$\left(\frac{\partial}{\partial t} - D\nabla^2\right)\rho_d = 0. \tag{4.2.3}$$

This is the original and the simplest diffusion equation. The diffusion equation can also be derived in other forms, but Eq. (4.2.2) is shown here to clarify its close relationship with the quantum mechanical flux, Eq. (3.1.27).

Now, when the Brownian particle is subject to an external field $V(q)$ in addition to the random force from the solvent and its motion is governed by the magnitude of $\rho_d(q, t)$, the time evolution of $\rho_d(q, t)$ is extended to

$$\frac{\partial}{\partial t}\rho_d(q,t) = -\vec{\nabla}\cdot\vec{j} - V(q)\rho_d \tag{4.2.4}$$

and thus, the diffusion equation becomes

$$\frac{\partial \rho_d}{\partial t} = \left(D\nabla^2 - V(q)\right)\rho_d. \tag{4.2.5}$$

This expression is to be compared with the Schrödinger equation later on.

4.3 Feynman-Kac Formula: Path Integral Over Random Paths and Integral Measure

4.3.1 Wiener Path Integral

For the case where the potential function $V(q)$ is absent, it is well known that the solution of the diffusion equation (4.2.3) after a time interval Δt is

$$\rho_d(q,\Delta t) = \frac{1}{(4\pi D\Delta t)^{1/2}}\exp\left[-\Delta t\left(\frac{1}{4D}\left(\frac{q-q_0}{\Delta t}\right)^2\right)\right], \tag{4.3.1}$$

with the initial condition (at $t = 0$) being

$$\rho_d(q,0) = \delta\left(q - q_0\right). \tag{4.3.2}$$

Since the Gaussian distribution function in Eq. (4.3.1) is nothing other than the Green function of the diffusion equation with the initial relation

$$\rho_d(q,0) = G_D(q,0), \tag{4.3.3}$$

G_D at any time t yields

$$G_D(q_1, t+\Delta t) = \frac{1}{(4\pi D\Delta t)^{1/2}} \times \int_{-\infty}^{\infty} dq_0 \exp\left[-\Delta t\left(\frac{1}{4D}\left(\frac{q_1-q_0}{\Delta t}\right)^2\right)\right] G_D(q_0,t). \tag{4.3.4}$$

Furthermore, continuing this process

$$G_D(q_2, t+2\Delta t) = \frac{1}{(4\pi D\Delta t)^{1/2}}\frac{1}{(4\pi D\Delta t)^{1/2}}\int_{-\infty}^{\infty} dq_1 \int_{-\infty}^{\infty} dq_0$$

$$\times \exp\left[-\Delta t\left(\frac{1}{4D}\left(\frac{q_2-q_1}{\Delta t}\right)^2\right)-\Delta t\left(\frac{1}{4D}\left(\frac{q_1-q_0}{\Delta t}\right)^2\right)\right]G_D(q_0,t) \tag{4.3.5}$$

shows that if a Brownian particle starts at point q_0 at time t, passes through point q_1 at time $t+\Delta t$, and then reaches point q_2 at time $t+2\Delta t$, the transition probability for such a path is thus sequentially given. In terms of formulas, q_1 and q_0, as well as q_2 and q_1, are connected by straight lines, but if Δt is taken sufficiently small and the sequence $q_0 \to q_1 \to q_2 \to q_3 \to q_4 \cdots \to q_n$ forms a continued line, then in $G_D(q_n, t+(n+1)\Delta t)$, the continuous curves approximately represented by many connected straight lines are realized. Therefore, if we multiply

$$\frac{1}{(4\pi D\Delta t)^{1/2}}dq_k \exp\left[-\Delta t\left(\frac{1}{4D}\left(\frac{q_{k+1}-q_k}{\Delta t}\right)^2\right)\right] \tag{4.3.6}$$

for each interval $q_k \to q_{k+1}$ and continue this process, we can construct infinitely many sets of global curves (paths), each to be summed up with a single weight (measure). In this way, the multiple integral in the form of Eq. (4.3.5) can be interpreted as summing over all possible curves that maintain continuity. This is just the path integral of Wiener [4].

4.3.2 Feynman-Kac Formula

The quantum mechanical path integral in real-valued space is discussed in detail in Chap. 6 from the viewpoint of sum over possible events or histories. For that purpose, it is necessary to extend Wiener's result to the case where there is a potential function $V(q)$ [3, 5, 6]. Letting the diffusion constant be D, the diffusion equation is given by Eq. (4.2.5):

$$\frac{\partial}{\partial t}\psi_D(q,t) = \left(D\nabla^2 - \lambda V(q,t)\right)\psi_D(q,t). \tag{4.3.7}$$

Hereafter, $\psi_D(q,t)$ is taken as the solution to the diffusion equation. For later convenience, we introduce a parameter λ. This path integral representation, as in the case of Brownian motion, in the limit $\Delta t \to 0$, becomes

$$\psi_D(x,t+\Delta t) = \frac{1}{(4\pi D\Delta t)^{1/2}}$$
$$\times \int_{-\infty}^{\infty} dy \exp\left[-\left(\frac{1}{4D}\frac{(x-y)^2}{\Delta t} + \lambda V(y,t)\Delta t\right)\right]\psi_D(y,t). \tag{4.3.8}$$

(This formula is used frequently, and the outline of its proof will be shown in the supplement below.) Thus, as above, the Green function can be written directly, but here we express it in the following form [3, 5, 6]:

$$G_D(q,t) = \lim_{N\to\infty} \frac{1}{(4\pi D\Delta t)^{N/2}} \prod_{k=1}^{N} \int_{-\infty}^{\infty} dq_k \times \exp\left[-\sum_{k=1}^{N} \Delta t \left(\frac{1}{4D}\left(\frac{\Delta q_k}{\Delta t}\right)^2 + \lambda V(q_k, t_k)\right)\right], \tag{4.3.9}$$

where $t_k = k\Delta t$, $q_k = q(t_k)$, $q_0 = 0$, $q_N = q$, and $\Delta q_k = q_k - q_{k-1}$, with the initial condition $G_D(q,0) = \delta(q)$. This formula can be rewritten as follows:

$$G_D(q,t) = \int_{\Omega(q,t:0,0)} \exp\left[-\lambda \int_0^t V(s, X_s(\omega))ds\right] dP_{W(q,t:0,0)}(\omega). \tag{4.3.10}$$

Here, $\Omega\,(q, t : 0, 0)$ denotes the set of all paths on the (q, t)-plane that start from $(0, 0)$ and reach (q, t). Each individual path is distinguished by the symbol ω, which is used in the sense of a sample from the ensemble of paths. In this notation, $dP_{W(q,t:0,0)}(\omega)$

$$dP_{W(q,t:0,0)}(\omega) = \lim_{N\to\infty} \frac{1}{(4\pi D\Delta t)^{N/2}} \exp\left[-\sum_{k=1}^{N} \Delta t \left(\frac{1}{4D}\left(\frac{\Delta q_k}{\Delta t}\right)^2\right)\right] \prod_{k=1}^{N} dq_k \tag{4.3.11}$$

is the Wiener measure associated with Brownian motion free of potential function. The integral in Eq. (4.3.10) asserts that

$$\int_{\Omega(q,t:0,0)} \cdots dP_{W(q,t:0,0)}(\omega) \tag{4.3.12}$$

represents the sum over all paths ω contained in $\Omega\,(q, t : 0, 0)$, expressed in the form of a Stieltjes integral with the integral measure $dP_{W(q,t:0,0)}(\omega)$. Equation (4.3.10) thus expresses the probability that each path, weighted according to its form, moves over the potential $V(s, X_s(\omega))$.

Equation (4.3.10) is called the Feynman-Kac formula [5], which was published in 1949. It is said that Kac was inspired by Feynman's lectures on quantum path integrals. On the other hand, Feynman's path integral was published in 1948 [7], but it is said that Feynman was not aware of Wiener's path integral. Feynman's path integral is one of the main topics of this book and will be discussed again in detail in Chap. 6.

4.3.2.1 Supplement: Proof of Eq. (4.3.8)

Equation (4.3.8) is not entirely self-evident, and since it is a formula also used in quantum path integrals later, let us briefly review it. First, consider the integral I with the normalization constants in Eq. (4.3.8) omitted:

$$I = \int_{-\infty}^{\infty} dy \exp\left[-\left(\frac{1}{4D}\frac{(x-y)^2}{\Delta t} + \lambda V(y,t)\Delta t\right)\right] \psi_D(y,t). \tag{4.3.13}$$

Here, by changing variables

$$z = y - x \tag{4.3.14}$$

and writing I_0 as

$$I_0 = \int_{-\infty}^{\infty} dz \exp\left[-\left(\frac{1}{4D}\frac{z^2}{\Delta t} + \lambda V(x+z,t)\Delta t\right)\right] \psi_D(x+z,t), \tag{4.3.15}$$

we expand the integrand with respect to z as follows:

$$-\lambda V(x+z,t)\Delta t \simeq -\lambda V(x,t)\Delta t - \lambda V'(x,t) z \Delta t + \cdots \tag{4.3.16}$$

and

$$\psi_D(x+z,t) = \psi_D(x,t) + \psi_D'(x,t) z + \frac{1}{2}\psi_D''(x,t) z^2 + \cdots. \tag{4.3.17}$$

Here, among the terms in $V'(z,t)$, the lowest order in Δt is

$$\begin{aligned} &\int_{-\infty}^{\infty} dz \exp\left[-\left(\frac{1}{4D}\frac{z^2}{\Delta t} + \lambda V'(z,t)\Delta t z\right)\right] \psi_D(x+z,t) \\ &\simeq \int_{-\infty}^{\infty} dz \exp\left[-\left(\frac{1}{4D}\frac{z^2}{\Delta t}\right)\right] \left(1 + \lambda V'(z,t)\,(\Delta t)\, z\right) \psi_D'(x,t) z \end{aligned} \tag{4.3.18}$$

and we define I_1 as

$$I_1 \equiv \int_{-\infty}^{\infty} dz \exp\left[-\left(\frac{1}{4D}\frac{z^2}{\Delta t}\right)\right] \left(\lambda V'(z,t)\,(\Delta t)\, z\right) \left(\psi_D'(x,t) z\right). \tag{4.3.19}$$

Since

$$\int_{-\infty}^{\infty} z^2 \exp(-az^2) = \sqrt{\frac{\pi}{4}} a^{-3/2}, \tag{4.3.20}$$

I_1 is of order $(\Delta t)^{3/2}(\Delta t)^1$, and even considering the normalization constant $(\Delta t)^{-1/2}$, it becomes $(\Delta t)^2$, so it can be neglected from the first order in Δt. Thus, the integral is simply reduced to

$$I_0 = \exp(-\lambda V(x,t)\Delta t)\int_{-\infty}^{\infty} dz \exp\left(-\frac{1}{4D}\frac{z^2}{\Delta t}\right) \times \left[\psi_D(x,t) + \psi_D'(x,t)z + \frac{1}{2}\psi_D''(x,t)z^2\right]. \tag{4.3.21}$$

It then follows that using

$$\frac{1}{(4\pi D\Delta t)^{1/2}}\int_{-\infty}^{\infty} dz \exp\left(-\frac{1}{4D}\frac{z^2}{\Delta t}\right)\left[\psi_D(x,t) + \frac{1}{2}\psi_D''(x,t)z^2\right] = \psi_D(x,t) + D\Delta t\psi_D''(x,t) \tag{4.3.22}$$

and

$$\exp(-\lambda V(x,t)\Delta t) \simeq 1 - \lambda V(x,t)\Delta t, \tag{4.3.23}$$

we obtain

$$\psi_D(x,t+\Delta t) \simeq (1-\lambda V(x,t)\Delta t)\left(\psi_D(x,t) + D\Delta t\psi_D''(x,t)\right) \tag{4.3.24}$$

or

$$\frac{\psi_D(x,t+\Delta t) - \psi_D(x,t)}{\Delta t} = D\psi_D''(x,t) - \lambda V(x,t)\psi_D(x,t) - \lambda V(x,t)D\psi_D''(x,t)(\Delta t) \tag{4.3.25}$$

as the first order term in Δt. And the limit $\Delta t \to 0$ after all gives

$$\frac{\partial}{\partial t}\psi_D = D\psi_D''(x,t) - \lambda V(x,t)\psi_D(x,t). \tag{4.3.26}$$

4.4 Ito Stochastic Differential Equation

Einstein's theory of the diffusion equation was subsequently developed into a comprehensive theory of equations determining distribution functions, including the Fokker-Planck equation [8, 9]. On the other hand, the Langevin approach was elevated to a mathematically rigorous theory of stochastic processes through the work of Ito, Levy, and other mathematicians and physicists [10, 11].

4.4.1 Extension of Simple Stochastic Processes

Let X_t denote the position of a pollen grain (Brownian particle) at time t. For a certain time interval including this moment, the observed change is denoted as dX_t. When this particle undergoes the simplest and most typical Brownian motion, it can be represented by the following Wiener process $W(t, \omega)$ [10, 11], that is

$$dX_t = dW(t, \omega). \tag{4.4.1}$$

The transition probability of the Wiener process is subject to the Gaussian process:

$$P(q_2, t + \Delta t; q_1, \Delta t) = \frac{1}{(4\pi D\Delta t)^{1/2}} \exp\left[-\Delta t \left(\frac{1}{4D}\left(\frac{q_2 - q_2}{\Delta t}\right)^2\right)\right], \tag{4.4.2}$$

where D is the diffusion coefficient. The ω in Eq. (4.4.1) is introduced to distinguish individual trajectories (samples) of Brownian motion. As can be seen from Eq. (4.4.2), $P(q_2, t + \Delta t; q_1, \Delta t)$ does not depend on t (Markov process), and it does not depend on the individuality of ω either. Therefore, the essential property of the stochastic differentiation is characterized by

$$dW(t, \omega)\, dW(t, \omega) = 2Ddt \tag{4.4.3}$$

and

$$\left\langle \frac{dW(t', \omega)}{dt'} \frac{dW(t, \omega)}{dt} \right\rangle = 2D\delta\left(t' - t\right). \tag{4.4.4}$$

Next, consider the case where the Brownian particle is also driven by an external field $V(q, t)$ that varies with time and position. In this case, the Brownian particle undergoes changes in position according to the velocity field (defined by $\alpha(q, t)$). Here, we generalize Eq. (4.4.1) as follows [11]:

$$dX_t = \alpha(X_t, t)dt + b(t, X_t)dW(t, \omega). \tag{4.4.5}$$

The meaning of this is intuitively easy to understand. First, considering the possibility that the diffusion itself is affected by the field, we introduce a function $b(t, X_t)$ to modify the diffusion term. Next, the effect of the change in position due to the external force acting on the Brownian particle is reflected in the product of the drift velocity $\alpha(X_t, t)$ and the infinitesimal time interval dt. Here in this chapter, we assume that the external field does not cause significant changes to the diffusion process itself, and simply set $b(t, X_t) = 1$. In fact, when we apply the stochastic process theory in the context of quantum theory, $b(t, X_t)$ is a constant determined by the spatially uniform quantum diffusion. That is

$$dX_t = \alpha(X_t, t)dt + dW(t, \omega) \tag{4.4.6}$$

Thus, the remaining task is to determine $\alpha(X_t, t)$. In the context of diffusion theory, $\alpha(X_t, t)$ is often called the drift term, but in this text, in relation to quantum theory, we will consistently refer to it as the velocity field coefficient.

4.4.2 Stochastic Process and Velocity Field

In the case where there is no external field $V(q, t)$, the diffusion equation, the Wiener path integral, and the Ito stochastic differential equation are directly connected. When an external field is present, the Feynman-Kac formula gives the Green function of the diffusion equation in place of the Wiener path integral, and these two turn out to be directly related. Likewise, the extended Ito stochastic differential equation (4.4.6) must be related to the diffusion equation and the Feynman-Kac formula. To realize explicitly, it is necessary to express the velocity field coefficient $\alpha(X_t, t)$ as

$$\alpha(X_t, t) = 2D\frac{\nabla\psi_D}{\psi_D}, \tag{4.4.7}$$

where ψ_D is the solution to the diffusion equation (4.3.7). Consequently, if the Ito equation is written as

$$dX_t = 2D\frac{\nabla\psi_D}{\psi_D}dt + dW(t, \omega) \tag{4.4.8}$$

then it is consistent with the diffusion equation and the Feynman-Kac formula. The proof of Eq. (4.4.8) is rather involved. It is outlined below, referring to the summary by Brownian Motion [3].

We start from Eq. (4.4.6) with unknown form of the drift coefficient $\alpha(X_t, t)$. In this stochastic process, we want to consider the transition of X_t over a small time interval Δt and the probability of that transition:

$$(t, X_t) \rightarrow (t + \Delta t, X_{t+\Delta t}). \tag{4.4.9}$$

First, the underlying Wiener process is reconstructed (reproduced) according to Eq. (4.4.6) such that

$$W_{t+\Delta t} - W_t = X_{t+\Delta t} - X_t - \alpha(X_t, t)\Delta t. \tag{4.4.10}$$

(Hereafter, $W(t, \omega)$ will be denoted as W_t for simplicity.) Using this Wiener process, we want to reexpress the process of Eq. (4.4.9). The probability density that a Brownian particle at point (X_t, t) will reach a neighborhood Δx of point $X_{t+\Delta t}$ at time $t + \Delta t$ is represented as

$$
\begin{aligned}
&P_{\Delta x}(\Delta t) \\
&= \frac{\Delta x}{(4\pi D\Delta t)^{1/2}} \exp\left[-\frac{(W_{t+\Delta t} - W_t)^2}{4D\Delta t}\right] \\
&= \frac{\Delta x}{(4\pi D\Delta t)^{1/2}} \exp\left[-\frac{(X_{t+\Delta t} - X_t)^2}{4D\Delta t}\right] \\
&\times \exp\left[\frac{1}{2D}\alpha(X_t, t)\,(X_{t+\Delta t} - X_t) - \frac{1}{4D}\alpha(X_t, t)^2 \Delta t\right],
\end{aligned}
\tag{4.4.11}
$$

with use of Eq. (4.4.10) and taking advantage of the fact that W_t is a Gaussian process. In this formula, the factor in the second equality

$$
\frac{1}{(4\pi D\Delta t)^{1/2}} \exp\left[-\frac{(X_{t+\Delta t} - X_t)^2}{4D\Delta t}\right] \tag{4.4.12}
$$

represents the probability that X_t will reach a neighborhood Δx of $X_{t+\Delta t}$ in a short time interval under only the Gaussian process. Therefore, the notation is a bit cumbersome, but the infinitesimal transition probability for each path ω (recall W_t is an abbreviation for $W(t, \omega)$) is defined as

$$
dP_W(\Delta t, \omega) = \frac{\Delta x}{(4\pi D\Delta t)^{1/2}} \exp\left[-\frac{(X_{t+\Delta t}(\omega) - X_t(\omega))^2}{4D\Delta t}\right]. \tag{4.4.13}
$$

Thus, the original $P_{\Delta x}(\Delta t)$ can be rewritten as

$$
\begin{aligned}
&P_{\Delta x}(\Delta t) \\
&= dP_W(\Delta t, \omega) \exp\left[\frac{\alpha(X_t, t)}{2D}(X_{t+\Delta t} - X_t) - \frac{\alpha(X_t, t)^2}{4D}\Delta t\right].
\end{aligned}
\tag{4.4.14}
$$

With this formula, we see that we are approaching the Feynman-Kac formula (4.3.10). Furthermore, the part summarized by the exponential function in Eq. (4.4.14) represents the motion modulated by the potential function $V(q, t)$ and should correspond to the part $\exp\left[-\lambda \int_0^t V(s, X_s(\omega))ds\right]$ in Eq. (4.3.10). Furthermore, since $P_{\Delta x}(\Delta t)$ is the probability density of reaching a very small region Δx, we may define $P_X(\Phi, T)$ to be the probability that the path enters a finite region Φ within a finite time T. Then, $P_X(\Phi, T)$ can be expressed explicitly as

$$
\begin{aligned}
P_X(\Phi, T) &= \int P_{\Delta x}(\Delta t) d\omega \\
&= \int_\Phi \exp\left[\int \frac{\alpha(X_t, t)}{2D} dX_t - \int \frac{\alpha(X_t, t)^2}{4D} dt\right] dP_W(\omega).
\end{aligned}
\tag{4.4.15}
$$

Next, let us look in detail at the exponential function part of Eq. (4.4.14). First, define a function $A(q, t)$ in

$$\frac{\partial A(q,t)}{\partial q} = \alpha(q,t). \tag{4.4.16}$$

$A(q, t)$ is also determined by the same stochastic process, so its Ito formula can be written as follows:

$$\begin{aligned} dA(X,t) &= \frac{\partial A}{\partial t}dt + \frac{\partial A}{\partial X}dX + \frac{\partial^2 A}{\partial X^2}Ddt \\ &= \left(\frac{\partial A}{\partial t} + D\frac{\partial^2 A}{\partial X^2}\right)dt + \alpha(X,t)dX. \end{aligned} \tag{4.4.17}$$

Here, the formula

$$dXdX = 2Ddt \tag{4.4.18}$$

has been used. Thus, integrating Eq. (4.4.17) gives

$$\begin{aligned} \int_0^T \alpha(x,t)dX &= A(T) - A(0) \\ &\quad - \int_0^T \left(\frac{\partial A}{\partial t} + D\frac{\partial^2 A}{\partial x^2}\right)dt \end{aligned} \tag{4.4.19}$$

Substituting this back into Eq. (4.4.15), we obtain

$$\begin{aligned} P_X(\Phi, T) &= \int_\Phi \exp\left[\frac{1}{2D}(A(T) - A(0))\right] \\ &\times \exp\left[-\int_0^T \frac{1}{4D}\left(2\frac{\partial A}{\partial t} + 2D\frac{\partial^2 A}{\partial X^2} + \left(\frac{\partial A}{\partial X}\right)^2\right)dt\right] \\ &\times dP_W(\omega), \end{aligned} \tag{4.4.20}$$

which is exactly in the form of the Feynman-Kac formula (4.3.9). To complete the relationship, it is only necessary to

$$2\frac{\partial A}{\partial t} + 2D\frac{\partial^2 A}{\partial q^2} + \left(\frac{\partial A}{\partial q}\right)^2 = 4\lambda V(q,t). \tag{4.4.21}$$

Here, by substituting the real function $\psi_b(q, t)$ in the form

$$\psi_b(q,t) = \exp\left(\frac{1}{2D}A(q,t)\right) \tag{4.4.22}$$

into the above, Eq. (4.4.21) becomes

$$\frac{\partial \psi_b(q,t)}{\partial t} = \left(-D\frac{\partial^2}{\partial q^2} + \lambda V(q,t)\right)\psi_b(q,t), \tag{4.4.23}$$

which is nothing but the backward diffusion equation [11]. From Eqs. (4.4.22) and (4.4.16), we finally obtain

$$\alpha(q,t) = 2D\frac{1}{\psi_b}\frac{\partial \psi_b}{\partial q}. \tag{4.4.24}$$

By substituting the variable transformation and initial conditions, it is possible to reach the forward diffusion equation:

$$\frac{\partial \psi_f(q,t)}{\partial t} = \left(D\frac{\partial^2}{\partial q^2} - \lambda V(q,t)\right)\psi_f(q,t) \tag{4.4.25}$$

without losing generality, and the form Eq. (4.4.7) is finally attained. Thus, even in the case where there is an external field $V(q,t)$, it was found that the stochastic differential equation (4.4.8) is consistent with both the diffusion equation and the Feynman-Kac formula.

4.4.3 Relationship Between the Parabolic Partial Differential Equation and the Stochastic Process

For an explanation on the relationship between the parabolic partial differential equation and the stochastic process, I briefly refer to Nagasawa's *A New Quantum Theory Based on Markov Processes* (Sogensha/Misuzu Shobo, 2012). Let $q(t,x,t,y)$ be the transition probability density function and define the space-time evolution related to the initial function $f(y)$ as

$$u(x,t) = \int q(t,x,t,y)f(y)dy. \tag{4.4.26}$$

Here, we define the following moments with respect to $q(t,x,t,y)$:

$$\lim_{h\downarrow 0}\frac{1}{h}\int q(t-h,x,t,y) = 0, \tag{4.4.27}$$

$$b(x,t)_i = \lim_{h\downarrow 0}\frac{1}{h}\int q(t-h,x,t,y)(y_i - x_i)dy, \tag{4.4.28}$$

$$\sigma^2(x,t)_{ij} = \lim_{h\downarrow 0}\frac{1}{h}\int q(t-h,x,t,y)(y_i - x_i)(y_j - x_j)dy. \tag{4.4.29}$$

By doing so, the following parabolic partial differential equation is established for $u(x, t)$ and the moments:

$$\hat{L}u \equiv \frac{\partial u}{\partial t} + \frac{1}{2}\sum_{i,j} \sigma^2(x,t)_{ij} \frac{\partial u^2}{\partial x_i \partial x_j} + \sum_i b(x,t)_i \frac{\partial u}{\partial x_i}$$
$$= 0, \tag{4.4.30}$$

which is one of the theorems by Ito.

Before proceeding further, let us recall Ito's stochastic differential calculus, which leads to the following equation:

$$df(X_t, t) = \frac{\partial f}{\partial t}dt + \sum_i \frac{\partial f}{\partial x_i} dX_{ti} + \frac{1}{2}\sum_{i,j} \frac{\partial f^2}{\partial x_i \partial x_j} dX_{ti} dX_{tj}. \tag{4.4.31}$$

Then, we move to the stochastic process. In this context, we define the probability measure $\mathbf{P}_{(x,s)}$ for the probability of a subset of the sample space as

$$u(x, s) = \mathbf{P}_{(x,s)}\left[f(X_b)\right] = \int q(s, x, b, y) f(y) dy, \tag{4.4.32}$$

in which $q(s, x, b, y)$ is the same as the transition probability density function in Eq. (4.4.26). Now, if X_t is the solution of the stochastic process

$$dX_t = \sigma(x, t) dB_t + b(X_t, t) dt, \tag{4.4.33}$$

then, we obtain Ito's formula:

$$df(X_t, t) = \hat{L} f(X_t, t) dt + \sum_{i,j} \sigma^2(x,t)_{ij} \frac{\partial f(X_t, t)}{\partial x_i} dB_{tj} \tag{4.4.34}$$

by applying Ito's calculus above. Since it holds that

$$\lim_{h \downarrow 0} \frac{1}{h} \mathbf{P}_{(s,x)}\left[df\right] = 0, \tag{4.4.35}$$

$$\lim_{h \downarrow 0} \frac{1}{h} \mathbf{P}_{(s,x)}\left[dB\right] = 0, \tag{4.4.36}$$

and

$$\lim_{h \downarrow 0} \frac{1}{h} \mathbf{P}_{(s,x)}\left[\int_s^{s+h} \hat{L} f(X_r, r) dr\right] = \hat{L} f(x, s), \tag{4.4.37}$$

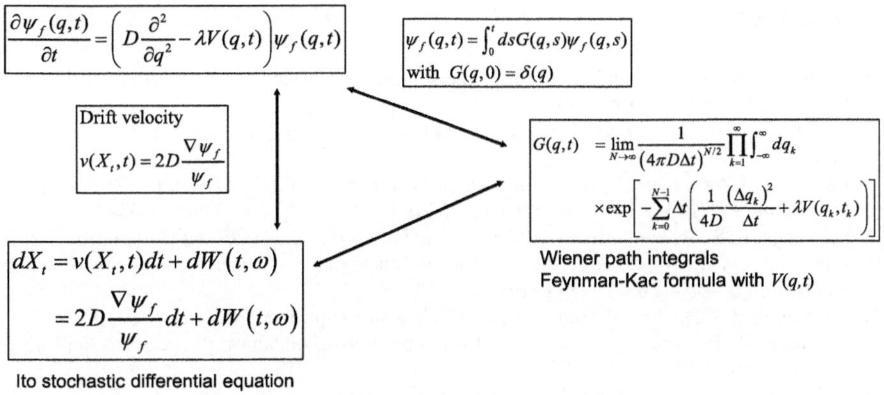

Fig. 4.1 Three pillars supporting the structure of a stochastic process

it is concluded in the sense of Eq. (4.4.30)

$$\hat{L} f = 0. \tag{4.4.38}$$

Since Eqs. (4.4.30) and (4.4.38) are equivalent, then, in summary, the Markov process subject to the stochastic differential equation (diffusion equation or the backward Fokker-Planck equation)

$$\frac{\partial u}{\partial t} + \frac{1}{2}\sigma^2 \nabla^2 u + b(q,t) \cdot \nabla u = 0 \tag{4.4.39}$$

is equivalent to the stochastic process:

$$X_t = X_0 + \sigma B_t + \int_0^t b(X_s, s) ds. \tag{4.4.40}$$

4.5 Three Pillar Structure in the Theory of Stochastic Process

Thus, the basic structure of the three important pillars in the physics of stochastic processes becomes clear: (1) the diffusion equation (4.3.7), (2) the Feynman-Kac path integral formula (4.3.10), and (3) Ito's stochastic differential equations (4.4.6) + (4.4.24). Since this structure is important, they are illustrated in Fig. 4.1 as a diagram.

References

1. Takatsuka, K. On the dual structure of the schrödinger dynamics. *Journal of the Physical Society of Japan* **94**, 094003 (2025).
2. Nagasawa, M. *Schrödinger Equations and Diffusion Theory* (Springer Science & Business Media, 2012).
3. Ezawa, H. & Nakamura, T. *Brownian Motion (in Japanese)*. Physics Mathematics Series, 336 p. (Asakura Publishing Co., Tokyo, Japan, 2020). ISBN 978-4-254-13792-7 C3342
4. Schulman, L. S. *Techniques and Applications of Path Integration* (Courier Corporation, 2012).
5. Kac, M. On distributions of certain Wiener functionals. *Transactions of the American Mathematical Society* **65**, 1–13 (1949).
6. Del Moral, P. & Del Moral, P. *Feynman-Kac Formulae* (Springer, 2004).
7. Feynman, R. P. Space-time approach to non-relativistic quantum mechanics. *Reviews of Modern Physics* **20**, 367 (1948).
8. Risken, H. *The Fokker-Planck Equation* (Springer, 1996).
9. Gardiner, C. *Handbook of Stochastic Methods for Physics, Chemistry and the Natural Sciences* (Springer, 2004).
10. Van Kampen, N. G. Stochastic differential equations. *Physics Reports* **24**, 171–228 (1976).
11. Oksendal, B. *Stochastic Differential Equations: An Introduction with Applications* (Springer Science & Business Media, 2013).

Real-Valued Schrödinger Equation: Rebuilding from Scratch

5

Abstract

We reconstruct the Schrödinger equation from scratch in the following logical structure:

1. The distribution density function of quantum particles $\rho(q,t)$ is decomposed as $\rho(q,t) = \bar{\psi}(q,t)^T \bar{\psi}(q,t)$, using a two-component real-valued Schrödinger vector $\bar{\psi}(q,t) = \begin{pmatrix} \phi_r(q,t) \\ \phi_c(q,t) \end{pmatrix}$. In quantum theory, $\rho(q,t)$ and $\bar{\psi}(q,t)$ play roles at two hierarchical levels and are complementary in this sense.
2. $\bar{\psi}(q,t)$ is the distribution amplitude function of a particle ensemble.
3. By imposing the continuity equation on $\rho(q,t)$ based on the Gauss theorem, and by imposing the space-time translational invariance, the structure of $\bar{\psi}(q,t)$ is established.
4. By applying the variational principle to the continuity equation for $\rho(q,t)$ with respect to $\bar{\psi}(q,t)$, the Schrödinger equation is obtained. The Schrödinger equation thus functions as a mathematical device that produces robust $\bar{\psi}(q,t)$ that has the continuity equation stationary.
5. $\bar{\psi}(q,t)$ is equivalent to the complex Schrödinger function. The imaginary unit in it corresponds to a geometrical rotation of the Schrödinger vector $\bar{\psi}(q,t)$.

Through the derivation, we find that the unitarity imposed on the Schrödinger vector makes quantum mechanics, while a semigroup space-time propagation brings it to classical diffusion dynamics.

New class of interpretations for the Schrödinger equation follows naturally and spontaneously.

K. Takatsuka, *Structure of the Schrödinger Dynamics*,
https://doi.org/10.1007/978-981-92-1910-0_5

Keywords

Real-valued Schrödinger equation · Schrödinger vector · Distribution amplitude function · Vector rotation for unitarity · Born's probabilistic interpretation · Bohr's complementarity · Unitary space-time propagation · Semigroup propagation operators · diffusion equation

Let us return to quantum theory. In this chapter, we rebuild the Schrödinger equation from first principles over the real number field. This equation is not a wave equation; accordingly, its solutions do not represent waves. Since it is not a wave-related function, we shall refrain, from this point onward, from using terms such as "wavefunction" or "wave equation." Instead, we simply employ the terms "Schrödinger function" and "Schrödinger equation."

Through a century of studies, we have come to "know" that the Schrödinger equation is "valid." In this chapter, we undertake to reconstruct (or re-derive) the Schrödinger equation from first principles. That said, we allow ourselves to rely, in part, on certain facts that are already established. For example, there is no need to re-examine the fact that quantities such as momentum and energy are expressed in units of the Planck constant. This is because the ultimate aim of this book is to clarify the physical meaning of the Schrödinger dynamics along with its underlying structure and its relationship to other theoretical frameworks, in particular diffusion theory. Relevant interpretation and understanding naturally follow from this analysis as consequences but are not artificially imported from independent schemes.

It is well known [1–4] that Erwin Schrödinger arrived at his equation in an essentially intuitive manner, often described as having "suddenly conceived" or "grasped" it at a fundamental level. As seen in Sect. 3.2 of Chap. 3, it is widely accepted that the emergence of the Schrödinger function is closely related to the solution surfaces of the Hamilton-Jacobi equation [5]. Throughout the history of quantum theory, various mathematical and physical frameworks have been proposed [1, 4]. In particular, the time-independent Schrödinger equation (3.2.18), being a relatively simple eigenvalue problem, has motivated attempts to reproduce it from alternative theoretical settings (a procedure commonly referred to as quantization).

Among the various quantization schemes, one of the most notable is the stochastic quantization proposed by Edward Nelson. This approach is based on a stochastic version of Newton's equation, from which the time-dependent Schrödinger equation can be derived through nonlinear transformations and a sequence of changes of variables [6, 7]. A brief introduction and discussion of this theory will be given in Chap. 8. According to Nelson, the origin of the Schrödinger equation lies in the stochastic Newton equation he introduced.

This naturally raises a fundamental question: What is the origin of his stochastic Newton equation itself? Is the stochastic Newton equation a more fundamental law than the Schrödinger equation is, or is it derived from a deeper underlying principle? Or, should the genuine Newton equation be derived as an approximate equation within quantum theory? For Einstein, who opposed Born's probability interpretation, the incorporation of stochastic processes into the fundamental dynamics law was not readily acceptable. Thus, after all, we are still puzzled about where we can seek further foundation leading to quantum mechanics.

Despite various criticisms, Nelson's attempt to derive the fundamental equation from more primitive assumptions is of considerable importance, and the author takes this perspective seriously, although not in full agreement with Nelson's conclusions.

In what follows, from this chapter onward, we develop a discussion of the Schrödinger dynamics, its origins, possible areas of future developments, and so on, all based on well-defined mathematical foundations.

5.1 Real-Valued Schrödinger Equation

In this chapter, we first clarify the origin of the Schrödinger equation and then examine its physical meaning on that basis.

5.1.1 Configuration Space Density Distribution Function

To begin, consider a finite system composed of electrons and nuclei, such as a molecule. We introduce a density distribution function $\rho(q,t)$, defined on configuration space, which reflects the motion of this ensemble of particles as a function of time. The function $\rho(q,t)$ is supposed to be positive semi-definite, $\rho(q,t) \geq 0$. Without loss of generality, it is assumed to be normalized such that

$$\int dq\rho(q,t) = 1. \tag{5.1.1}$$

Let us compare this with two classical distribution functions: the phase space density function $\Gamma(q,p,t)$, which is a solution of the classical Liouville equation (3.2.34), and the distribution function $\rho_d(q,t)$, which is a solution of the diffusion equation (4.2.3). Even in quantum theory, we can adopt a picture in which a collection of particles, each governed by some underlying dynamics, in space-time. In this sense, quantum distribution function $\rho(q,t)$ is conceptually analogous to the spirit of $\Gamma(q,p,t)$ and $\rho_d(q,t)$. Differences are however that $\Gamma(q,p,t)$ is defined in classical phase space, whereas $\rho_d(q,t)$ represents stochastic processes such as Brownian motion. $\rho(q,t)$ provides a direct description only in q-space under the constraints imposed by the uncertainty principle (yet see Eqs. (3.2.34) and (3.2.37)). Both of $\Gamma(q,p,t)$ and $\rho_d(q,t)$ give area-conserving maps, each satisfying the equation of continuity in its respective space: The Liouville equation (3.2.34) for $\Gamma(q,p,t)$, while the diffusion continuity equation (4.2.1) for $\rho_d(q,t)$.

It should be emphasized that none of $\Gamma(q,p,t)$, $\rho_d(q,t)$ and $\rho(q,t)$ do represent physical entities themselves, but rather a mathematical representation of the distributions of an ensemble of physical events. In most mathematical contexts, a distribution function does not represent a single realized phenomenon (event).

On the other hand, in order to track a single trajectory of an individual particle, a Brownian particle (a classical particle) for instance, we need to integrate the Langevin equation or stochastic differential equations (the Newtonian equation). Such an individual trajectory should be consistent with the distribution functions $\rho_d(q,t)$ ($\Gamma(q,p,t)$ for classical trajectories). Similarly, in quantum theory, if an equation governing individual dynamics comes to be available, it should be naturally consistent with the distribution function $\rho(q,t)$.

5.1.2 Vector Square Root of $\rho(q,t)$ and Its Dynamics: Schrödinger Vector

Since $\rho(q,t)$ is positive semi-definite, its square root as a real number always exists. Here, yet, we dare to consider the following square root:

$$\rho(q,t) = \left(\phi_r(q,t)\ \phi_c(q,t)\right)\begin{pmatrix}\phi_r(q,t)\\ \phi_c(q,t)\end{pmatrix} = \bar{\psi}(q,t)^T\bar{\psi}(q,t). \tag{5.1.2}$$

Here, the square root vector $\bar{\psi}(q,t)$, which we call the Schrödinger vector (not to be confused with the Dirac state vector), is

$$\bar{\psi}(q,t) \equiv \begin{pmatrix}\phi_r(q,t)\\ \phi_c(q,t)\end{pmatrix}. \tag{5.1.3}$$

The component functions $\phi_r(q,t)$ and $\phi_c(q,t)$ are real-valued functions and can take negative values as well. Furthermore, since $\rho(q,t)$ is a quadratic form, any orthogonal transformation (vector rotation) that preserves it can exist, so $\bar{\psi}(q,t)$ is not uniquely determined in that sense. In other words, one vector corresponds to only one ρ, but one ρ can correspond to infinitely many vectors. Therefore, once we promote $\bar{\psi}(q,t)$ classically up to $\rho(q,t)$, it generally cannot be uniquely mapped back to the same quantum state.

By introducing the above square root in terms of the Schrödinger vector, we have brought additional dimensionality into the dynamics space behind $\rho(q,t)$ from N to 2N–1, for N-dimensional q. In Eq. (5.1.2), we used a two-component vector, but it is not limited to two-component vectors. For instance

$$\bar{\psi}(q,t) \equiv \begin{pmatrix}\phi_a(q,t)\\ \phi_b(q,t)\\ \phi_c(q,t)\\ \phi_d(q,t)\end{pmatrix} \tag{5.1.4}$$

is equally accepted to make a square root of $\rho(q,t)$. (In this case, it is likely for quaternion to play the main role in place of the complex numbers.) In fact, in relativistic quantum theory, more components are generally required. However, as will be shown later, the nonrelativistic Schrödinger mechanics happens to be closed within the two-component vector scheme.

Under the above circumstances, let us consider the following fundamental conditions for $\rho(q,t)$ in nonrelativistic quantum scheme and see how they are reflected at the level of the Schrödinger vector.

Condition 1 The translational symmetry (invariance) of the space q where $\rho(q,t)$ is defined. Here, we assume a simple Cartesian coordinate system.

Condition 2 The translational symmetry (invariance) of the time axis. One may arbitrarily set the origin of time.

Condition 3 The non-compressibility of $\rho(q,t)$ and the equation of continuity equation associated with it.

It is clear that these conditions are not specific only to quantum theory. What is unique is that the conditions are applied to the Schrödinger vector $\bar{\psi}(q,t)$. On the other hand, it is obvious that the translational invariance of Euclidean space-time does not hold in general relativity or the Big Bang theory. Let us examine below whether the ordinary Schrödinger equation can be derived under these conditions.

5.1.3 Translational Invariance in q-Space of no External Field

Let us consider a parallel translation of the origin of the q coordinate of the Schrödinger vector, $q \to q + \Delta q$. From the appearance, $\bar{\psi}(q,t)$ seems to change only as a function representation

$$\bar{\psi}(q,t) \to \bar{\psi}(q+\Delta q,t). \tag{5.1.5}$$

That is, it transforms as such. In this case, since $\bar{\psi}(q,t)$ also has the degree of freedom of vector rotation, the transformation to $\bar{\psi}(q+\Delta q,t)$ may also involve some rotation. Therefore, the general expression for the transformation can be written as

$$\begin{aligned}\bar{\psi}(q+\Delta q,t) &= \exp\left(\Delta q \mathbf{A}\vec{\nabla}_q\right)\bar{\psi}(q,t)\\ &= \left(\mathbf{I} + \Delta q \mathbf{A}\vec{\nabla}_q\right)\begin{pmatrix}\phi_r(q,t)\\ \phi_c(q,t)\end{pmatrix} + (\text{higher}),\end{aligned} \tag{5.1.6}$$

In this formula, the matrix $\mathbf{A}$ in the evolution operator $\exp\left(\Delta q \mathbf{A}\vec{\nabla}_q\right)$ is a real 2×2 matrix, and in the second row in Eq. (5.1.6), it is expanded to the first order. The transposed vector of this formula is

$$\tilde{\psi}(q+\Delta q,t)^T = \left(\phi_r(q,t)\ \phi_c(q,t)\right)\exp\left(\Delta q \mathbf{A}^T \vec{\nabla}_q\right). \tag{5.1.7}$$

Hereafter, the superscript T on the matrix denotes the transpose. Of course, the case where $\mathbf{A}$ is the identity matrix is also included. However, as will be seen later, this vector rotation makes a definite difference of quantum dynamics from the Brownian motion.

As seen in Formula (5.1.6), $\mathbf{A}\vec{\nabla}_q$ serves as a shift operator. Along with this, the condition for unitarity of $\rho(q,t)$

$$\tilde{\psi}(q+\Delta q,t)^T\tilde{\psi}(q+\Delta q,t) = \bar{\psi}(q,t)^T\bar{\psi}(q,t) \tag{5.1.8}$$

gives

$$\begin{aligned}
&\tilde{\psi}(q+\Delta q,t)^T\tilde{\psi}(q+\Delta q,t) - \bar{\psi}(q,t)^T\bar{\psi}(q,t)\\
&= \left(\phi_r(q,t)\ \phi_c(q,t)\right)\left[\mathbf{A}^T\vec{\nabla}_q + \mathbf{A}\vec{\nabla}_q\right]\begin{pmatrix}\phi_r(q,t)\\ \phi_c(q,t)\end{pmatrix}\Delta q\\
&= 0,
\end{aligned} \tag{5.1.9}$$

which requires that $\mathbf{A}$ must satisfy

$$\mathbf{A}^T = -\mathbf{A}. \tag{5.1.10}$$

(Here, we have shown it in the first-order expansion, but this generally holds). In the case of a 2×2 real matrix, we have no choice other than

$$\mathbf{A} = c_p\mathbf{J}. \tag{5.1.11}$$

(In Eq. (5.1.9), the Hermiticity of the differential operator $\vec{\nabla}_q$ may be of concern. However, since we are dealing with two identical real vectors, it is sufficient to consider the transposition symmetry of the matrix $\mathbf{A}$.) Here, $\mathbf{J}$ is the familiar standard symplectic matrix:

$$\mathbf{J}=\begin{pmatrix}0 & -1\\ 1 & 0\end{pmatrix}, \tag{5.1.12}$$

and c_p is any real constant. It is well known that $\mathbf{J}$ is the fundamental element that characterizes the symplectic structure in classical mechanics [8, 9], and it already appeared in Eq. (3.2.4).

The constant c_p cannot be determined within the framework of the present theory and ultimately should be chosen so as to reproduce the experimental values. However, some preparation is necessary to proceed. According to Noether's theorem and its spirit, the shift symmetry of spatial translation corresponds to the conservation of momentum. So, as Dirac did [10], we define the quantum mechanical momentum operator $\hat{p}$ using the shift operator as follows:

$$\hat{p} = c_p \mathbf{J} \vec{\nabla}_q . \tag{5.1.13}$$

Incidentally, in Eq. (5.1.9), it is naturally understood that "the quantum mechanical operator should be placed to act between two Schrödinger vectors," and as shown on the right side of Eq. (5.1.9), to calculate the momentum, we write in such a manner:

$$\left(\phi_r(q,t)\ \phi_c(q,t) \right) \hat{p} \begin{pmatrix} \phi_r(q,t) \\ \phi_c(q,t) \end{pmatrix} . \tag{5.1.14}$$

(It is neither obvious nor trivial to convince the beginners in quantum theory why the expectation values are taken in such a manner as in Eq. (5.1.14).) In Dirac's notation [10], the expectation value of an operator $\hat{A}$ for a state vector, say, $|\psi_1\rangle$, is represented as $\left\langle \psi_1 \left| \hat{A} \right| \psi_1 \right\rangle$, and the theoretical framework is beautifully axiomized as such.

Here, let us relate c_p to the conventional quantum theory: We begin with a pure state $\rho(q,t)$ defined as follows, for simplicity

$$\rho(q,t) = 1.0 \text{ in } q \in [q_1(t), q_2(t)] . \tag{5.1.15}$$

We assume $q_i(t) = q_i(t_0) + p_0(t - t_0)/m$ $(i = 1, 2)$. The length of the interval $[q_1(t), q_2(t)]$ is assumed to be sufficiently long. $\rho(q,t)$ is 0 outside the range. m is the mass. This $\rho(q,t)$ is moving with a constant velocity p_0/m. In this special example, since $\phi_r^2(q,t) + \phi_c^2(q,t) = \rho(q,t) = 1.0$, without losing generality, we may set $\bar{\psi}(q,t)$ as

$$\phi_r(q,t) = \cos(\frac{p_0}{K} q + \chi) \quad \text{and} \quad \phi_c(q,t) = \sin(\frac{p_0}{K} q + \chi) \tag{5.1.16}$$

within the region $q \in [q_1(t), q_2(t)]$ and zero elsewhere. Here, K and χ are constants. If we may choose $[q_1(t), q_2(t)] = [-\infty, \infty]$, we observe

$$\hat{p} \begin{pmatrix} \phi_r(q,t) \\ \phi_c(q,t) \end{pmatrix} = c_p \frac{p_0}{K} \begin{pmatrix} \phi_r(q,t) \\ \phi_c(q,t) \end{pmatrix} . \tag{5.1.17}$$

But due to the effect of the edges of the range, $\begin{pmatrix} \phi_r(q,t) \\ \phi_c(q,t) \end{pmatrix}$ cannot be an eigenfunction of $\hat{p}$. Anyway, we obtain

$$\left(\phi_r(q,t)\ \phi_c(q,t) \right) \hat{p} \begin{pmatrix} \phi_r(q,t) \\ \phi_c(q,t) \end{pmatrix} = -\frac{p_0}{K} c_p . \tag{5.1.18}$$

in the interval. Now, since this value is expected to coincide with p_0, we expect

$$-\frac{p_0}{K}c_p = p_0, \tag{5.1.19}$$

which thereby gives $c_p = -K$. On the other hand, in order for the present formulation to be consistent with experimental facts [11], it is necessary to set

$$K = \hbar, \tag{5.1.20}$$

which is in parallel with what we did in Eq. (3.2.19).

Thus, in real-valued Schrödinger dynamics, the momentum operator turns out to be given as

$$\hat{p} = -\hbar \mathbf{J}\vec{\nabla}. \tag{5.1.21}$$

5.1.4 Translational Invariance in Time

The treatment of time translational invariance is fundamentally the same as in the spatial case. The time-shifted Schrödinger vector $\tilde{\psi}(q, t + \Delta t)$ is written as

$$\tilde{\psi}(q, t + \Delta t)^T = \exp(\Delta t \mathbf{B}\partial_t)\,\bar{\psi}(q, t) \tag{5.1.22}$$

and

$$\tilde{\psi}(q, t + \Delta t)^T = \left(\phi_r(q, t)\ \phi_c(q, t)\right)\exp\left(\Delta t \mathbf{B}^T \partial_t\right), \tag{5.1.23}$$

and thereby the time-shift (evolution) operator (matrix) is $\mathbf{B}\partial_t$. Under this operation, $\rho(q, t)$ remains unchanged, as should be expressed as

$$\tilde{\psi}(q, t + \Delta t)^T \tilde{\psi}(q, t + \Delta t) = \bar{\psi}(q, t)^T \bar{\psi}(q, t). \tag{5.1.24}$$

Thus, similarly

$$\mathbf{B} = c_t \mathbf{J} \tag{5.1.25}$$

is required. Time-reversal symmetry in this context is reflected in the law of conservation of energy. Hence, the shift operator is formally expressed using the Hamiltonian $\hat{H}$ as

$$\hat{H} = c_t \mathbf{J}\frac{\partial}{\partial t} \tag{5.1.26}$$

Here, we also take the constant c_t, which will be determined later, in advance and define it as

$$c_t = \hbar.$$

(Here again, c_t has to be determined to reproduce the experimental facts.) The explicit form of the Hamiltonian $\hat{H}$ will be determined later, but up to this point, we have so far obtained

$$\begin{pmatrix} \phi_r & \phi_c \end{pmatrix} \left(\hbar \mathbf{J} \frac{\partial}{\partial t} \right) \begin{pmatrix} \phi_r \\ \phi_c \end{pmatrix} = \begin{pmatrix} \phi_r & \phi_c \end{pmatrix} \hat{H} \begin{pmatrix} \phi_r \\ \phi_c \end{pmatrix} \tag{5.1.27}$$

as a natural result.

5.1.5 Continuity Equation for $\rho(q,t) = \bar{\psi}(q,t)^T \bar{\psi}(q,t)$

$\rho(q,t)$ preserves its volume in q-space and should satisfy the equation of continuity:

$$\frac{\partial}{\partial t}\rho(q,t) = -\nabla \cdot \vec{j}(q,t), \tag{5.1.28}$$

where $\vec{j}(q,t)$ is a relevant flux that we want to identify. We remember, in the classical diffusion process, that the flux is given by $\vec{j}(q,t) = -D\vec{\nabla}\rho_d(q,t)$ in Eq. (4.2.2), which in turn leads directly to the diffusion equation. On the other hand, in classical fluid dynamics, the flux is simply expressed as

$$\vec{j}(q,t) = \vec{v}(q,t)\rho_d(q,t), \tag{5.1.29}$$

where $\rho(q,t)$ is transported by the velocity field $\vec{v}(q,t)$.

In quantum theory, the flux of the flow is not necessarily separated into the velocity field $\vec{v}(q,t)$ and $\rho(q,t)$. Meanwhile, the quantity that should be called the "momentum density" is given by Eq. (5.1.14). Therefore, it is natural to divide this by the mass m to obtain the velocity density and then adopt this as the quantum-theoretical flux. That is

$$\begin{aligned} \vec{j}(q,t) &\equiv \begin{pmatrix} \phi_r(q,t) & \phi_c(q,t) \end{pmatrix} \frac{\hat{p}}{m} \begin{pmatrix} \phi_r(q,t) \\ \phi_c(q,t) \end{pmatrix} \\ &= \frac{\hbar}{m} \left(\phi_r \vec{\nabla}\phi_c - \phi_c \vec{\nabla}\phi_r \right). \end{aligned} \tag{5.1.30}$$

By substituting this into the continuity equation (5.1.28), we obtain

$$\begin{aligned} \begin{pmatrix} \phi_r & \phi_c \end{pmatrix} \frac{\partial}{\partial t} \begin{pmatrix} \phi_r \\ \phi_c \end{pmatrix} &= -\frac{\hbar}{2m} \vec{\nabla} \cdot \left(\phi_r \vec{\nabla}\phi_c - \phi_c \vec{\nabla}\phi_r \right) \\ &= -\frac{\hbar}{2m} \begin{pmatrix} \phi_r & \phi_c \end{pmatrix} \vec{\nabla}^2 \begin{pmatrix} \phi_c \\ -\phi_r \end{pmatrix} \\ &= \begin{pmatrix} \phi_r & \phi_c \end{pmatrix} \left(-\mathbf{J} \frac{\hat{p}^2}{2m} \right) \begin{pmatrix} \phi_r \\ \phi_c \end{pmatrix}. \end{aligned} \tag{5.1.31}$$

This expression gives a better perspective by rewriting to

$$(\phi_r\ \phi_c)\,\hbar\mathbf{J}\frac{\partial}{\partial t}\begin{pmatrix}\phi_r\\ \phi_c\end{pmatrix} = (\phi_r\ \phi_c)\left(\frac{\hat{p}^2}{2m}\right)\begin{pmatrix}\phi_r\\ \phi_c\end{pmatrix}, \tag{5.1.32}$$

where

$$\hat{p}^2 = -\hbar^2\mathbf{J}^2\nabla^2 = \hbar^2\nabla^2\mathbf{I}. \tag{5.1.33}$$

with $\mathbf{I}$ being the unit matrix.

5.1.6 Inclusion of Potential Terms

Equation (5.1.32) is still far from the Schrödinger equation in real-valued space, simply because the potential function $V(q,t)$ is missing. In order to incorporate $V(q,t)$, we recall the skew orthogonality, which was already used in Eq. (3.2.6), which is

$$(\phi_r\ \phi_c)\,\mathbf{J}\begin{pmatrix}\phi_r\\ \phi_c\end{pmatrix} = 0. \tag{5.1.34}$$

This, of course, holds for any scalar function:

$$(\phi_r\ \phi_c)\,\mathbf{J}f(q,t)\begin{pmatrix}\phi_r\\ \phi_c\end{pmatrix} = 0. \tag{5.1.35}$$

We here take

$$f(q,t) = V(q,t), \tag{5.1.36}$$

and coming back to Eq. (5.1.31) in such a manner

$$\begin{aligned}(\phi_r\ \phi_c)\left(\hbar\frac{\partial}{\partial t}\right)\begin{pmatrix}\phi_r\\ \phi_c\end{pmatrix}\\ = (\phi_r\ \phi_c)\left(-\mathbf{J}\frac{\hat{p}^2}{2m} - \mathbf{J}V(q,t)\right)\begin{pmatrix}\phi_r\\ \phi_c\end{pmatrix}\end{aligned} \tag{5.1.37}$$

without loss of generality. The reason for the choice of Eq. (5.1.36) is that we naturally come to the Hamiltonian:

$$\hat{H} = \frac{\hat{p}^2}{2m} + V(q,t). \tag{5.1.38}$$

After all, Eq. (5.1.37) turns out to be

$$\begin{aligned}&\left(\phi_r\ \phi_c\right)\left(\hbar\mathbf{J}\frac{\partial}{\partial t}\right)\begin{pmatrix}\phi_r\\ \phi_c\end{pmatrix}\\ &=\left(\phi_r\ \phi_c\right)\left(\frac{\hat{p}^2}{2m}+V(q,t)\right)\begin{pmatrix}\phi_r\\ \phi_c\end{pmatrix}.\end{aligned} \tag{5.1.39}$$

We here recall that $V(q,t)$ has already disappeared from the quantum mechanical equation of continuity (3.1.27). It disappeared because we considered the complex conjugate of the Schrödinger equation and took the sum of the two. Therefore, the potential function $V(q,t)$ became missing in the stage of the equation of continuity.

We also recall that the diffusion equation for $\rho_d(q,t)$ could be derived from the equation of continuity (4.2.1) by simply inserting in it a natural (classical) form of the flux, Eq. (4.2.2). In quantum mechanics, however, we are seeking for the dynamical equation for $\psi(q,t)$ rather than $\rho(q,t)$, and hence, we need one more step to proceed.

5.1.7 Derivation of the Schrödinger Equation via the Variational Principle

The form of Eq. (5.1.39), which appears to be the prototype of the Schrödinger equation, could be derived by imposing only a few conditions, so that infinitely many solutions can exist. To understand this point, let us return to the continuity equation, Eq. (5.1.28). In general, in fields such as fluid dynamics and electromagnetism, the continuity equation is used to determine how the distribution density function $\rho(q,t)$ evolves under given initial and boundary conditions, once the flow field—namely, the flux $\vec{j}(q,t)$—is specified. In contrast, in Eq. (5.1.39), both $\rho(q,t)$ and $\vec{j}(q,t)$ are constructed from the same underlying functions $\begin{pmatrix}\phi_r\\ \phi_c\end{pmatrix}$, and therefore, the equation is not of the type in which one specifies $\vec{j}(q,t)$ and solves for $\rho(q,t)$. Rather, it merely prescribes the relationship between the quantities $\begin{pmatrix}\phi_r\\ \phi_c\end{pmatrix}$ appearing on both sides of Eq. (5.1.39). In other words, Eq. (5.1.39) is not an equation of motion but simply an expression of a conservation law, and consequently, there exist infinitely many pairs of ϕ_r and ϕ_c that satisfy it. Accordingly, our objective is to single out a particular class from among this continuous infinity of possible $\begin{pmatrix}\phi_r\\ \phi_c\end{pmatrix}$.

Now, since Eq. (5.1.39) represents a functional of the Schrödinger vector in a quadratic form, it is natural to conceive that the $\begin{pmatrix}\phi_r\\ \phi_c\end{pmatrix}$ that makes the functional stationary with respect to its first order variation should be selected. The stationarity implies that such a solution is robust against a small variation of the Schrödinger vector. Such a Schrödinger vector is to be picked with the variational principle (sometimes referred to as the Dirac-Frenkel variational principle) as

$$\begin{aligned}&\left(\delta\phi_r\ \delta\phi_c\right)\left(\hbar\frac{\partial}{\partial t}\right)\begin{pmatrix}\phi_r\\ \phi_c\end{pmatrix}\\ &=\left(\delta\phi_r\ \delta\phi_c\right)\mathbf{J}\left(\frac{\hat{p}^2}{2m}+V(q,t)\right)\begin{pmatrix}\phi_r\\ \phi_c\end{pmatrix}=0.\end{aligned} \tag{5.1.40}$$

The right-hand side variation

$$\begin{aligned}&\left(\phi_r\ \phi_c\right)\left(\hbar\frac{\partial}{\partial t}\right)\begin{pmatrix}\delta\phi_r\\ \delta\phi_c\end{pmatrix}\\ &=\left(\phi_r\ \phi_c\right)\mathbf{J}\left(\frac{\hat{p}^2}{2m}+V(q,t)\right)\begin{pmatrix}\delta\phi_r\\ \delta\phi_c\end{pmatrix}=0\end{aligned} \tag{5.1.41}$$

is also required independently. Here, note that the skew orthogonality in (5.1.35) has lost its validity, that is

$$\left(\delta\phi_r\ \delta\phi_c\right)\mathbf{J}V(q,t)\begin{pmatrix}\phi_r\\ \phi_c\end{pmatrix}\neq 0. \tag{5.1.42}$$

Now, for the first time, $V(q,t)$ begins to possess the mathematical role to dominate dynamics.

Since $\left(\delta\phi_r\ \delta\phi_c\right)$ can be arbitrary, we may extract as usual:

$$\hbar\frac{\partial}{\partial t}\begin{pmatrix}\phi_r\\ \phi_c\end{pmatrix}=-\mathbf{J}\hat{H}\begin{pmatrix}\phi_r\\ \phi_c\end{pmatrix}, \tag{5.1.43}$$

which is equivalent to

$$\begin{aligned}\hbar\mathbf{J}\frac{\partial}{\partial t}\begin{pmatrix}\phi_r\\ \phi_c\end{pmatrix}&=\hat{H}\begin{pmatrix}\phi_r\\ \phi_c\end{pmatrix}\\ &=\left(-\frac{\hbar^2}{2m}\nabla^2+V\right)\begin{pmatrix}\phi_r\\ \phi_c\end{pmatrix}.\end{aligned} \tag{5.1.44}$$

This is the real-valued Schrödinger-type equation.

$\hat{H}$ and $\hat{p}$ are respectively summarized once again as

$$\hat{H}=\mathbf{J}\hbar\frac{\partial}{\partial t} \tag{5.1.45}$$

and

$$\hat{p} = -\mathbf{J}\hbar\vec{\nabla}, \tag{5.1.46}$$

both of which are supposed to be operated on the Schrödinger vectors.

5.1.8 Role and Function of the Schrödinger Equation

The real-valued Schrödinger equation obtained here will be shown in the next section to be equivalent to the original complex Schrödinger equation formulated by Schrödinger. At this point, the reader should note that Eq. (5.1.44) has been derived in accordance with very fundamental physical requirements and a natural conceptual framework. By expressing the square root of the distribution density function $\rho(q, t)$ in vector form, the resulting equation takes a form that does appear neither in classical mechanics nor in stochastic dynamics. The Schrödinger vector $\bar{\psi}(q, t)$ in Eq. (5.1.3) can take negative values, and yet, it possesses properties analogous to those of a distribution function within the algebraic structure at the level of quantum mechanics. Anticipating this, it is appropriate to refer to $\bar{\psi}(q, t)$ as a distribution amplitude function.

It should be emphasized here again that the Schrödinger vector $\bar{\psi}(q, t)$ forms a quadratic form in the continuity equation (5.1.39) and that the Schrödinger equation is expected to work "to select the most robust state from the continuity equation" or "to choose the vector function that satisfies the equation of continuity with the highest stability." This is precisely the physical meaning and role of the Schrödinger equation.

Incidentally, the fact that variational principles play an essential role is by no means unusual in the natural sciences. However, it may be stated that the variational principle employed here (Eq. (5.1.40)) is far more transparent in meaning than the Maupertuis variational principle—the principle of least action—which underlies classical mechanics. (It is said that Maupertuis, when asked why his variational principle holds, attributed it to "the perfection of divine wisdom.")

In an effort to move beyond the circular definitions between the Schrödinger equation and the Schrödinger function inherent in the Copenhagen interpretation—namely, that "the Schrödinger (wave) function evolves according to the Schrödinger equation when not observed, and that the Schrödinger equation serves to govern the time evolution of wavefunction"—various alternative interpretations have been proposed. The present derivation of the real-valued Schrödinger equation from first principles has cut off the loop in the circular definitions. And thereby, it is now expected that some of the rather unnatural and unnecessary assumptions and additional interpretations external to the formalism, which effectively ascribe to the Schrödinger equation properties that it does not inherently possess—such as wavefunction collapse and the wavefunction of the universe—could be eventually abandoned.

5.1.9 Complex Schrödinger Equation

Here, let us confirm that the real-valued Schrödinger equation (5.1.44) is equivalent to the original Schrödinger equation. Recall the symplectic unit vector **J** is naturally introduced in (5.1.6) and (5.1.22) in order to preserve the unitarity of the distribution density function $\rho(q, t)$ with respect to the parallel transport in space-time. The most characteristic property of **J** is

$$\mathbf{J}^2 = -\mathbf{I}, \tag{5.1.47}$$

and

$$\mathbf{J}^{-1} = -\mathbf{J} = \mathbf{J}^T \tag{5.1.48}$$

so that, for real vectors, it has the same effect as the imaginary unit i, and thus, mathematically, they are almost equivalent under the replacement that

$$\mathbf{J} \rightarrow i. \tag{5.1.49}$$

As such, the two fundamental operators in Eq. (5.1.45) and Eq. (5.1.46) can be rewritten over the field of complex numbers, respectively, as

$$\hat{H} = i\hbar \frac{\partial}{\partial t} \tag{5.1.50}$$

and

$$\hat{p} = -i\hbar\vec{\nabla}. \tag{5.1.51}$$

By comparing the Hamiltonian of Eqs. (5.1.50) and (5.1.45) and also the momentum operator of Eqs. (5.1.46) and (5.1.51), we summarize that the differential operators ∂_t and $\vec{\nabla}$ arose as a generator for translational shift operation in time and space, respectively, while the imaginary unit i has emerged from the vector rotation.

The Schrödinger vector $\bar{\psi}(q, t) = \begin{pmatrix} \phi_r \\ \phi_c \end{pmatrix}$ can directly read as the Schrödinger function:

$$\psi(q, t) = \phi_r(q, t) + i\phi_c(q, t) \tag{5.1.52}$$

and by identifying $\bar{\psi}(q, t)$ and $\psi(q, t)$, we can treat them as the same. In this way, Eq. (5.1.44) becomes identical to the complex Schrödinger equation:

$$i\hbar \frac{\partial}{\partial t}\psi(q, t) = \left(-\frac{\hbar^2}{2m}\nabla^2 + V\right)\psi(q, t) \tag{5.1.53}$$

In fact, after substituting Eq. (5.1.52) into Eq. (5.1.53) and separating into real and imaginary parts, it becomes clear that it directly leads back to Eq. (5.1.44).

Given his extraordinary insight, Schrödinger was able to arrive directly at Eq. (5.1.53), entering at once into the innermost structure of quantum theory. However, the presence of imaginary numbers may be likened to the discomfort of wearing leather shoes on a Japanese tatami-mat floor; this incongruity manifested itself in the formalism and led to interpretations such as a mechanics of wave. In this book, we instead construct a proper entrance, proceed straightforwardly along the corridor, and enter a "real-valued room" adjacent to Schrödinger's inner chamber. The real-valued room is connected to the domain of real-valued path integrals (Chap. 6) and further leads to a hidden domain—rarely explored—of quantum stochastic differential equations (Chap. 8).

In this way, we understand why complex numbers appear in the Schrödinger equation despite its status as one of the most fundamental equations. We also comprehend that the Schrödinger equation stands only in the four-dimensional simple Euclidean space and the reason why the spatial derivative in it is closed within the second order.

Nevertheless, this does not imply that practical calculations in quantum mechanics should be reformulated in terms of the real-valued Schrödinger equation. The complex Schrödinger equation remains both convenient and elegant. Although not pursued in this book, it is evident that adherence to real numbers is not advantageous when developing relativistic theories or quantum field theory. However, in the present text, we continue to investigate the origin and meaning of the Schrödinger dynamics through the real-valued path integrals as Green's functions and quantum stochastic differential equations. Accordingly, unless otherwise stated, our discussion will be made within the framework of the real-valued Schrödinger equation in what follows.

5.2 Some Properties of the Real-Valued Schrödinger Equation

5.2.1 Heisenberg Dynamics

Quantum mechanics has a structure of operator algebra. As the very first step, it is possible to derive the Heisenberg equation of motion from the real-valued Schrödinger equation. Since we will use this later, let us briefly explain the essentials here. An operator $\hat{A}$ that is assumed not to depend on time satisfies the following equation:

$$
\begin{aligned}
&\frac{\partial}{\partial t}\left(\phi_r(q,t)\ \phi_c(q,t)\right)\hat{A}\begin{pmatrix}\phi_r(q,t)\\ \phi_c(q,t)\end{pmatrix}\\
&= -\frac{1}{\hbar}\left(\phi_r\ \phi_c\right)\left(\mathbf{J}^T\hat{H}\hat{A}+\hat{A}\hat{H}\mathbf{J}\right)\begin{pmatrix}\phi_r\\ \phi_c\end{pmatrix}\\
&\equiv \frac{1}{\hbar}\left(\phi_r\ \phi_c\right)\left[\hat{H},\hat{A}\right]\mathbf{J}\begin{pmatrix}\phi_r\\ \phi_c\end{pmatrix}.
\end{aligned}
\tag{5.2.1}
$$

Here, the commutator is defined as usual by $\left[\hat{H}, \hat{A}\right] = \hat{H}\hat{A} - \hat{A}\hat{H}$ with the Hamiltonian $\hat{H}$. These act on a Schrödinger vector as a matrix operator with $\mathbf{J}$. Also, it is assumed that $\hat{H}$ is a Hermitian evolution operator acting on a two-component vector $\begin{pmatrix} \phi_r \\ \phi_c \end{pmatrix}$. When the evolution operator $\hat{A}$ depends explicitly on time, $\left[\hat{H}, \hat{A}\right]$ in Eq. (5.2.1) can simply be replaced by $\mathbf{J}$ in

$$\left[\hat{H}, \hat{A}\right]\mathbf{J} + \partial\hat{A}(t)/\partial t. \tag{5.2.2}$$

Among others, if we set $\hat{A} = \hat{H}$, it becomes clear that the expectation value of the energy is conserved. Furthermore, the Heisenberg representation of an operator can be formulated in a similar manner to that in the complex Schrödinger dynamics.

5.2.2 Time-Reversal Equation

Since the Schrödinger vector does not have its complex conjugate, a little care is necessary in considering the time-reversal. First, prepare the following matrix:

$$\mathbf{P}_3 = \begin{pmatrix} 1 & 0 \\ 0 & -1 \end{pmatrix}, \tag{5.2.3}$$

and operate it from the left side of Eq. (5.1.44) as

$$\mathbf{P}_3 \hbar \mathbf{J} \frac{\partial}{\partial t} \begin{pmatrix} \phi_r \\ \phi_c \end{pmatrix} = \mathbf{P}_3 \left(-\frac{\hat{p}^2}{2m} + V(q,t) \right) \begin{pmatrix} \phi_r \\ \phi_c \end{pmatrix}. \tag{5.2.4}$$

$\mathbf{P}_3$ and $\mathbf{J}$ mutually anti-commute as

$$\mathbf{P}_3\mathbf{J} = -\mathbf{J}\mathbf{P}_3, \tag{5.2.5}$$

and the following transformation follows:

$$-\hbar \mathbf{J} \frac{\partial}{\partial t} \mathbf{P}_3 \begin{pmatrix} \phi_r \\ \phi_c \end{pmatrix} = \left(-\frac{\hat{p}^2}{2m} + V(q,t) \right) \mathbf{P}_3 \begin{pmatrix} \phi_r \\ \phi_c \end{pmatrix}, \tag{5.2.6}$$

which also reads

$$-\hbar \mathbf{J} \frac{\partial}{\partial t} \begin{pmatrix} \phi_r(q,t) \\ -\phi_c(q,t) \end{pmatrix} = \left(-\frac{\hat{p}^2}{2m} + V(q,t) \right) \begin{pmatrix} \phi_r(q,t) \\ -\phi_c(q,t) \end{pmatrix}. \tag{5.2.7}$$

Then, replacing $t \rightarrow -t$, we obtain the time-reversed real Schrödinger equation such that

$$\hbar \mathbf{J} \frac{\partial}{\partial t} \begin{pmatrix} \phi_r(q,-t) \\ -\phi_c(q,-t) \end{pmatrix} = \left(-\frac{\hat{p}^2}{2m} + V(q,-t)\right) \begin{pmatrix} \phi_r(q,-t) \\ -\phi_c(q,-t) \end{pmatrix}. \tag{5.2.8}$$

It is interesting to compare this result with the time-reversal of a classical trajectory $(q(t), p(t))$, say $(q_{rev}(t), p_{rev}(t))$, with $q_{rev}(-t) = q(t)$ and $p_{rev}(t) = -p(-t)$. Equation (5.2.8) seems rather complicated, but it is actually equivalent to

$$i\hbar \frac{\partial}{\partial t} \psi^*(q,-t) = \hat{H}\psi^*(q,-t). \tag{5.2.9}$$

In this way, the unitarity of the Schrödinger equation is confirmed.

Let us recall that in the complex Schrödinger dynamics, the density $\rho(q,t)$ can be written as

$$\rho(q,t) = \psi^*(q,t)\psi(q,t) = \psi(q,-t)\psi(q,t), \tag{5.2.10}$$

which causes a little confusion about the physical meaning of $\rho(q,t)$. In the real Schrödinger dynamics, $\rho(q,t)$ is merely

$$\rho(q,t) = \left(\phi_r(q,t)\ \phi_c(q,t)\right) \begin{pmatrix} \phi_r(q,t) \\ \phi_c(q,t) \end{pmatrix} = \bar{\psi}(q,t)^T \bar{\psi}(q,t), \tag{5.2.11}$$

and no time-reversal argument is necessary to define $\rho(q,t)$ at current time, unlike in the time-symmetric interpretation [12].

5.2.3 Velocity Field and Energy Field Generated by the Rotation of the Schrödinger Vector

5.2.3.1 Space-time Distribution of Energy and Momentum

The Schrödinger vector can also be normalized by the usual method:

$$\int dq \bar{\psi}(q,t)^T \bar{\psi}(q,t) = \int dq \rho(q,t) = 1. \tag{5.2.12}$$

Likewise, the expectation value of the energy expectation value can be obtained as

$$E^{\mathrm{av}} = \left\langle \hat{H} \right\rangle = \frac{\int dq \left(\phi_r(q,t)\ \phi_c(q,t)\right) \hat{H} \begin{pmatrix} \phi_r(q,t) \\ \phi_c(q,t) \end{pmatrix}}{\int \rho(q,t) dq}. \tag{5.2.13}$$

We may also perform the local normalization of energy, giving the spatiotemporal distribution of energy as

$$
\begin{aligned}
E^{\text{local}}(q,t) &= \frac{1}{\rho}\left(\phi_r(q,t)\ \phi_c(q,t)\right)\mathbf{J}\hbar\frac{\partial}{\partial t}\begin{pmatrix}\phi_r(q,t)\\ \phi_c(q,t)\end{pmatrix} \\
&= -\frac{\hbar}{\rho}\left(\phi_r\partial_t\phi_c - \phi_c\partial_t\phi_r\right) \\
&= \text{Re}\,\frac{\hat{H}\psi}{\psi}.
\end{aligned}
\tag{5.2.14}
$$

Note that even if $E^{\text{local}}(q,t)$ is integrated over q, it does not necessarily coincide with E^{av}, simply because the density ρ appears in the denominator of Eq. (5.2.14).

Similarly, the spatiotemporal distribution of momentum can be defined by using $\hat{p} = -\mathbf{J}\hbar\vec{\nabla}$ as

$$
\begin{aligned}
\vec{p}^{\text{local}}(q,t) &= \frac{1}{\rho}\left(\phi_r(q,t)\ \phi_c(q,t)\right)(-\mathbf{J}\hbar\vec{\nabla})\begin{pmatrix}\phi_r(q,t)\\ \phi_c(q,t)\end{pmatrix} \\
&= \frac{\hbar}{\rho}\left(\phi_r\vec{\nabla}\phi_c - \phi_c\vec{\nabla}\phi_r\right) \\
&= \text{Re}\,\frac{\vec{p}\psi}{\psi} = \hbar\,\text{Im}\,\frac{\vec{\nabla}\psi}{\psi}.
\end{aligned}
\tag{5.2.15}
$$

The last term in this expression plays an important role as the velocity field that drives the particle in the quantum path stochastic differential equation discussed in Chap. 8. The average value of the local momentum $\vec{p}^{\text{local}}(q,t)$ is

$$
\begin{aligned}
\int \rho(q,t)\vec{p}^{\text{local}}(q,t)dq &= \int dq\hbar\left(\phi_r\vec{\nabla}\phi_c - \phi_c\vec{\nabla}\phi_r\right) \\
&= \int dq\left(\phi_r(q,t)\ \phi_c(q,t)\right)\hat{p}\begin{pmatrix}\phi_r(q,t)\\ \phi_c(q,t)\end{pmatrix} \\
&= \langle\vec{p}\rangle \qquad \text{(not normalized)},
\end{aligned}
\tag{5.2.16}
$$

which clearly demonstrate the meaning that $\vec{p}^{\text{local}}(q,t)$ is the momentum density at space-time (q,t).

5.2.3.2 Energy and Momentum in the Polar Coordinate Representation

By expressing the Schrödinger vector $\bar{\psi}(q,t)$ in the polar coordinate representation of a vector in such a way that

$$
\bar{\psi}(q,t) = \rho(q,t)^{1/2}\begin{pmatrix}\cos\theta(q,t)\\ \sin\theta(q,t)\end{pmatrix},
\tag{5.2.17}
$$

or equivalently for the complex Schrödinger function

$$\psi(q,t) = \rho(q,t)^{1/2} \exp\left(i\theta(q,t)\right), \tag{5.2.18}$$

both $\vec{p}^{\text{local}}(q,t)$ and $E^{\text{local}}(q,t)$ are rewritten in a simpler form. The representation of Eq. (5.2.18) is nothing other than the Bohm representation in Eq. (3.2.20) [13–16]. Using Eq. (5.2.17), $\vec{p}^{\text{local}}(q,t)$ and $E^{\text{local}}(q,t)$ are expressed, respectively, as

$$\vec{p}^{\text{local}}(q,t) = \hbar\vec{\nabla}\theta(q,t) \tag{5.2.19}$$

and

$$E^{\text{local}}(q,t) = -\hbar\partial_t\theta(q,t). \tag{5.2.20}$$

which are only of the functions of the angle coordinates $\theta(q,t)$.

These representations clearly indicate that the rotation of the Schrödinger vector plays an essential role in quantum dynamics. At the same time, they show that the function $\theta(q,t)$ contains an indeterminacy up to an additive constant. The spatiotemporal distribution of momentum can be expressed as the spatial derivative (gradient) of the rotation of the Schrödinger vector, while that of energy is proportional to the rate of the rotation. A distribution point located at q is transported to a neighboring point with velocity $\vec{\nabla}\theta(q,t)$, and the variation of energy at (q,t) is governed by the time derivative of $\theta(q,t)$. Thus, from the standpoint of the distribution function, each point in space-time is endowed with a rotation of the Schrödinger vector, and the internal rotational dynamics drive the quantum distribution dynamics.

From Eqs. (5.2.19) and (5.2.20), it is natural to infer that the function $\hbar\theta(q,t)$ plays the role of an action integral in quantum theory. This is precisely the viewpoint discussed by Bohm in connection with Eq. (3.2.20) and the subsequent discussion. Conversely, from the standpoint of the real Schrödinger vector, the emergence of Bohm's representation can be explained in a transparent manner. In contrast, it is presumably difficult to derive the Schrödinger equation starting from Bohm's formulation. From the perspective of the Schrödinger vector, the Schrödinger equation can be derived straightforwardly as shown above, and also, Bohm's representation naturally appears in the form of a polar coordinate representation.

Incidentally, when $\theta(q,t)$ satisfies

$$\partial_t^2\theta(q,t) = 0, \tag{5.2.21}$$

a stationary state arises, satisfying

$$E^{\text{local}}(q,t) = -\hbar\partial_t\theta(q,t) = E \quad \text{(constant)} \tag{5.2.22}$$

This is, of course, equivalent to the separation $\psi(q,t) = \Phi(q)\exp\left(-iEt/\hbar\right)$, which leads to

$$\hat{H}\Phi(q) = E\Phi(q). \tag{5.2.23}$$

As a side remark, it is known that the positions of discrete energy spectra in quantum Hamiltonian systems, namely, the eigenenergies, can be obtained using only the quantum phase even in chaotic regimes. Since $\theta(q,t)$ corresponds to the quantum phase, this approach is referred to as phase quantization [17–19].

5.2.4 Semigroup Time Evolution for Diffusion Equation

By the way, the reader may have noticed that in Eq. (5.2.19) for $\vec{p}^{\text{local}}(q,t)$ and in Eq. (5.2.20) for $E^{\text{local}}(q,t)$, neither

$$\vec{\nabla}\rho(q,t) \tag{5.2.24}$$

nor

$$\partial_t \rho(q,t) \tag{5.2.25}$$

did appear. In particular, we recall that $\vec{\nabla}\rho(q,t)$ plays a decisive role in diffusion phenomena, as can be clearly seen in Eq. (4.2.2). The mathematical reason why this does not appear is that in the calculations of $\vec{p}^{\text{local}}(q,t)$ and $E^{\text{local}}(q,t)$, the term

$$(\cos\theta(q,t) \quad \sin\theta(q,t))\,\mathbf{J}\begin{pmatrix}\cos\theta(q,t)\\ \sin\theta(q,t)\end{pmatrix} \tag{5.2.26}$$

vanishes due to the skew orthogonality by $\mathbf{J}$ (see Eq. (5.1.34)). Here is hidden an important quantum mechanical property.

To investigate this aspect, we define the quantity $p^{\text{Imag}}(q,t)$, which is obtained by removing $\mathbf{J}$ from Eq. (5.2.15), as follows:

$$\vec{p}^{\text{Imag}}(q,t) \equiv \frac{1}{\rho}\left(\phi_r(q,t)\ \phi_c(q,t)\right)(-\hbar\vec{\nabla})\begin{pmatrix}\phi_r(q,t)\\ \phi_c(q,t)\end{pmatrix}. \tag{5.2.27}$$

This time, it holds

$$\left(\cos\theta(q,t)\ \sin\theta(q,t)\right)\vec{\nabla}\begin{pmatrix}\cos\theta(q,t)\\ \sin\theta(q,t)\end{pmatrix} = 0, \tag{5.2.28}$$

so the term $\vec{\nabla}\theta(q,t)$ disappears, resulting in

$$p^{\text{Imag}}(q,t) = -\frac{\hbar}{2\rho}\left(\vec{\nabla}\rho\right). \tag{5.2.29}$$

With use of the complex Schrödinger function ψ, we may re-express $p^{\text{Imag}}(q,t)$ as represented as

$$p^{\text{Imag}}(q,t) \equiv \text{Im}\,\frac{\vec{p}\psi}{\psi} = -\frac{\hbar}{2\rho}\left(\vec{\nabla}\rho\right), \tag{5.2.30}$$

and exactly corresponds to the imaginary part of $\vec{p}\psi/\psi$ (compare with Eq. (5.2.15)). This form gives another key to suggest the similarity and difference between quantum mechanics and diffusion theory.

Furthermore, from Eq. (5.2.30), we see

$$-\frac{\hbar}{2m}\left(\vec{\nabla}\rho\right) = \frac{p^{\text{Imag}}(q,t)}{m}\rho, \tag{5.2.31}$$

which corresponds to the classical flux (velocity distribution). Comparing this flux and that of the diffusion theory, Eq. (4.2.2), that is

$$\vec{j} = -D\vec{\nabla}\rho_d, \tag{5.2.32}$$

it gives rise to

$$D = \frac{\hbar}{2m} \tag{5.2.33}$$

(see Eq. (3.2.33)).

Regarding the energy expectation value, the same thing holds. The form of the space-time distribution of energy with the complex Schrödinger function is

$$\begin{aligned}\frac{\hat{H}\psi}{\psi} &= \frac{1}{\rho}\left(\phi_r\left(-\frac{\hbar^2}{2m}\nabla^2+V\right)\phi_r+\phi_c\left(-\frac{\hbar^2}{2m}\nabla^2+V\right)\phi_c\right)\\ &+\frac{i}{\rho}\left(-\frac{\hbar^2}{2m}\right)\left(\phi_r\nabla^2\phi_c-\phi_c\nabla^2\phi_r\right)\\ &\equiv E^{\text{local}}(q,t)+iE^{\text{Imag}}(q,t).\end{aligned} \tag{5.2.34}$$

so, taking the real part, we have

$$E^{\text{local}}(q,t) = -\hbar\partial_t\theta(q,t) \tag{5.2.35}$$

(with reference to Eqs. (5.2.4) and (5.2.20)) for the imaginary part. On the other hand, by applying the continuity equation (5.1.28), it is possible to write as

$$E^{\text{Imag}}(q,t) = \frac{\hbar}{2\rho}\frac{\partial\rho}{\partial t}. \tag{5.2.36}$$

Thus, $E^{\text{Imag}}(q,t)$ is not an energy distribution in quantum theory. In this way, when considering the non-quantum mechanical dynamical variables $p^{\text{Imag}}(q,t)$ and energy $E^{\text{Imag}}(q,t)$, it becomes clear that, in their respective operators, the symplectic unit matrix $\mathbf{J}$ has been replaced by the unit matrix $\mathbf{I}$, i.e., $\mathbf{J} \to \mathbf{I}$. As a result, $\hat{p}$ and $\hat{H}$ become

$$\hat{p} = -\mathbf{J}\hbar\vec{\nabla} \;\to\; \hat{p}^{\text{Imag}} = -\hbar\mathbf{I}\vec{\nabla} \tag{5.2.37}$$

and

$$\hat{H} = \mathbf{J}\hbar\frac{\partial}{\partial t} \;\to\; \hat{H}^{\text{Imag}} = \hbar\mathbf{I}\frac{\partial}{\partial t}, \tag{5.2.38}$$

respectively, and are replaced by the non-quantum momentum $\hat{p}^{\text{Imag}}$ and the non-quantum energy $\hat{H}^{\text{Imag}}$.

In this way, in the world of the elementary Schrödinger equation, the "i" that appeared to be just a number actually plays an important role as the symplectic vector rotation through $\mathbf{J}$, which distinguishes between unitarity and the semigroup in the time evolution. Thus, we are now able to answer the fundamental question in quantum theory, "Why does the Schrödinger equation contain an imaginary number?" The non-quantum mechanical dynamical variable $p^{\text{Imag}}(q,t)$ will be discussed later in Chap. 8 in the context of the quantum stochastic differential equation.

Let us add a note just in case: A semigroup is a structure obtained from a group by removing the requirement for the existence of inverses. Here, while the quantum time evolution operators are forming a unitary group with parameters t and q, in response to this, in the case of a semigroup, for the time evolution operator $\hat{T}(t)$, it suffices to require only

$$\lim_{t\to+0}\hat{T}(t) = I, \qquad \hat{T}(t)\hat{T}(s) = \hat{T}(t+s) \tag{5.2.39}$$

instead.

5.2.5 On $\rho(q,t)$ and $\bar{\psi}(q,t)$

In light of the derivation of the Schrödinger equation presented above, we conclude this chapter with supplementary remarks on the Schrödinger function and the distribution density function.

5.2.5.1 Complementarity Between $\rho(q,t)$ as a Classical Distribution and $\bar{\psi}(q,t)$ as a Quantum Distribution

At a time when Schrödinger had suddenly introduced his equation and the complex function $\psi(q,t)$ and when neither their physical meaning nor their origin was well understood, Born's proposal of a probabilistic interpretation of $\psi^*(q,t)\psi(q,t)$ was truly revolutionary. Einstein's famous reaction—"God does not play dice" (1926) [4]—was no doubt rooted in his strong belief in deterministic natural laws, but it was also likely influenced by the lack of a firm mathematical foundation for the probabilistic interpretation. Indeed, one reason for this strong resistance may have been that the complex function $\psi(q,t)$ and the quantity $\psi^*(q,t)\psi(q,t)$ can be regarded as independent or fundamentally distinct objects. The often-cited statement that "$\psi^*(q,t)\psi(q,t)$ represents the probability density of finding the system in a given state" may therefore have appeared too great a conceptual leap as an interpretation of $\psi(q,t)$.

In the present book, however, we have proceeded from a different starting point. We assume that a distribution density function $\rho(q,t)$ exists a priori, and we construct its square-root representation in terms of a real-valued vector function $\bar{\psi}(q,t)$, such that $\rho(q,t) = \bar{\psi}(q,t)^T\bar{\psi}(q,t)$. In this formulation, the direct relationship between $\rho(q,t)$ and $\bar{\psi}(q,t)$ is manifest at the outset. More precisely, the function $\bar{\psi}(q,t)$, which may be regarded as the vectorial square root of the distribution density function $\rho(q,t)$ that is observable at the classical level, plays the role of a distribution amplitude function. It encodes the causal structure (i.e., the dynamical law in the quantum sense) at the quantum level. The dynamics of $\bar{\psi}(q,t)$ are governed by the Schrödinger equation, which satisfies space-time translational invariance and makes the continuity equation associated with particle number conservation stationary.

Thus, it is evident that $\rho(q,t)$ and $\bar{\psi}(q,t)$ (or $\psi(q,t)$) are related through a direct and complementary mathematical relationship. However, it should be emphasized that a "distribution density function" is not identical to a "probability density function" in the usual probabilistic interpretation. If the distribution is realized only probabilistically, then there must exist an underlying mathematical and physical basis for such characterization. In fact, in addition to the constraint arising from quantum uncertainty—which prevents the complete and unique reproducibility of experimental conditions—there exists another essential stochastic process underlying the Schrödinger equation itself. This will be examined in Chap. 8.

While $\rho(q,t)$ is a quantity we can recognize in classical world, $\bar{\psi}(q,t)$ and $\psi(q,t)$ represent quantum dynamics as a distribution amplitude and are not necessarily directly observable. Any attempt to observe it may disturb the system, potentially altering even the original $\rho(q,t)$. This leads to a dual structure of epistemic limitation. Why does such a structure arise? The functions $\bar{\psi}(q,t)$ or $\psi(q,t)$ exist and evolve—even when unobserved—much like the Moon, yet to access the outcome of that evolution, one must rely on $\rho(q,t)$. The concept of complementarity introduced by Niels Bohr is often interpreted as implying that human understanding of nature requires two mutually complementary descriptions. However, the relationship between the worlds of $\bar{\psi}(q,t)$ and $\rho(q,t)$ is not merely bridged by Born's probability interpretation; rather, they are directly connected through the square root relation, with $\bar{\psi}(q,t)$ being the vectorial square root of the positive semi-definite function $\rho(q,t)$. According to the standpoint of this book, the origin of physical probabilistic behavior must therefore lie elsewhere, and will be discussed in Chaps. 8 and 9.

5.2.5.2 The Schrödinger Functions Represent the Distribution of Ensemble of Events

A distribution function, by its very nature, is of a mathematical structure that represents a large number of events within a single functional form. This point was already emphasized in connection with the classical Liouville function in the paragraph following Eq. (3.2.34). The same applies to $\rho_d(q, t)$ in diffusion processes and $\rho(q, t)$ in quantum theory: A single function contains an uncountably large number of possible events.

Moreover, in quantum theory, since $\bar{\psi}(q, t)$ is the "square root" of $\rho(q, t)$, it represents a coherent sum over many such events. While this may appear natural once the origin and meaning of $\bar{\psi}(q, t)$ are properly understood, misunderstanding can arise otherwise. In particular, one may mistakenly regard a single $\bar{\psi}(q, t)$ as representing a single event. Expressions such as the "wavefunction of the universe" or statements that, in the double-slit experiment, "a particle passes through two slits simultaneously and interferes with itself" are based on such misconceptions. Although the Schrödinger function can pass through two slits simultaneously, such a nonphysical interpretation would not be tenable if it were to represent a single electron.

The most intuitive framework illustrating that the Schrödinger function represents many possible events is the path integral formulation. However, since the path integral in the sense of Feynman involves additional mathematical subtleties, it will be examined in greater depth in the next chapter from the standpoint of the real-valued Schrödinger equation.

5.2.5.3 The Schrödinger Function Does Not Collapse

Both the distribution density function $\rho(q, t)$ and the distribution amplitude functions $\bar{\psi}(q, t)$ and $\psi(q, t)$ are mathematical entities. They reflect properties of matter but do not constitute direct representations of matter itself. Consequently, there is no basis to introduce any notion of physical "collapse" into these functions.

This situation is analogous to Brownian motion: Observing the position of a Brownian particle under a microscope does not cause the corresponding diffusion distribution function $\rho_d(q, t)$ to collapse to a point. In quantum mechanics, while the distribution amplitude function can be logically tracked by the Schrödinger equation up to the moment of observation, the act of measurement may introduce a significant perturbation not accounted for in the prior dynamics, thereby decisively affecting subsequent evolution.

In the double-slit experiment, the appearance of localized spots on a detection screen—evidence of electron impacts—may tempt one to conclude that "something associated with the electron has collapsed at that point" [1–4]. However, what physically represents the arrival event will be examined in Chap. 8. At the very least, the Schrödinger equation predicts the distribution $\rho(q, t)$ for the possible point of impact, but it does not determine which outcome is realized—only that probabilities are proportional to the magnitude of $\rho_d(q, t)$. This is just a mathematical process.

Physically, in the double-slit experiment, an electron collides with a molecule on the detection screen, transferring energy in electron recombination reactions that results in emission of light, which is then amplified and recorded as a signal. There is no need to invoke further mystery. If one wishes to include the collision process within the Schrödinger framework, it can be treated within quantum scattering theory, including emission processes. Even in that case, however, it is merely a classical picture to assume that the electron remains localized at the point of impact. As often observed in quantum scattering theory, a portion of the Schrödinger function may become spatially localized, but this is fundamentally different from the instantaneous collapse of the entire function to a single point.

Similarly, when a quantum particle such as an electron encounters a potential barrier like the detection panel, a wave packet (e.g., a Gaussian packet) may appear to contract sharply along the direction of propagation before being reflected. In semiclassical approximations such as the WKB method, this process may yield divergent solutions at that point. Although this may resemble a collapse, it is not. First, such apparent divergences can be regularized into finite-width functions by quantum smoothing [20]. Second, even if contraction seems to occur along the propagation direction (normal to the detection plane), the wave packet does not become sharply localized in directions parallel to the panel.

For these reasons, the author rejects the notion of wave function collapse as proposed in the Copenhagen interpretation and further suggests that many alternative theories having been proposed thus far to avoid collapse may be unnecessary.

5.2.5.4 Wave-Like Behavior of the Schrödinger Function and Matter Waves

The principle of complementarity proposed by Niels Bohr originated from reflections on how to understand the duality of particle-like and wave-like behavior. While electrons appear as particles in classical descriptions, their wave-like behavior in quantum theory is undeniable: Experiments designed to detect particle properties reveal particle-like behavior, whereas those designed to detect wave properties reveal wave-like behavior.

In Sect. 3.1, the following questions were raised regarding matter waves:

(1) Are the so-called matter waves intrinsic properties of matter?
(2) Does the Schrödinger function directly represent matter waves?

Phenomena such as diffraction patterns observed in electron transmission through thin crystals (as in the experiments of Davisson, Germer, and Kikuchi) and interference patterns in double-slit experiments (e.g., by Tonomura) are often interpreted as interference of "physical matter waves." However, as clarified in this chapter, the Schrödinger vector and function are not physical waves but distribution amplitude functions; they do not directly represent a physical wave. Indeed, it is now widely accepted that the Schrödinger function does not correspond to a physically real wave [1–4]. Consequently, there is currently little basis for affirming that matter waves are intrinsic properties of matter. Nevertheless, because the Schrödinger equation is linear, it naturally exhibits wave-like features such as superposition and interference. The origin of interference patterns in single-particle double-slit experiments, however, remains to be clarified and will be discussed in Chap. 8.

References

1. Auletta, G. *Foundations and Interpretation of Quantum Mechanics: In the Light of a Critical-Historical Analysis of the Problems and of a Synthesis of the Results* (World Scientific, 2001).
2. Home, D. & Whitaker, A. *Einstein's Struggles with Quantum Theory: A Reappraisal* (Springer Science & Business Media, 2007).
3. Whitaker, A. *The New Quantum Age: from Bell's Theorem to Quantum Computation and Teleportation* (Oxford University Press, 2012).
4. Freire Jr, O. *et al. The Oxford Handbook of the History of Quantum Interpretations* (Oxford University Press, 2022).
5. Simeonov, L. S. Derivation of Schrodinger's equation from the Hamilton-Jacobi and the Eikonal equations. *arXiv preprint arXiv:2410.03689* (2024).
6. Nelson, E. Derivation of the Schrödinger equation from newtonian mechanics. *Physical Review* **150**, 1079 (1966).
7. Nelson, E. Review of stochastic mechanics. *Journal of Physics: Conference Series*, vol. 361, 012011 (IOP Publishing, 2012).
8. Arnold, V. I. *Mathematical Methods of Classical Mechanics* (Springer, Berlin, 1989), 2 edn.
9. De Gosson, M. A. *Symplectic Geometry and Quantum Mechanics*, vol. 166 (Springer Science & Business Media, 2006).
10. Dirac, P. A. M. *The Principles of Quantum Mechanics* (Oxford University Press, 1981).
11. Messiah, A. *Quantum Mechanics* (Courier Corporation, 2014).
12. Aharonov, Y., Popescu, S. & Tollaksen, J. A time-symmetric formulation of quantum mechanics. *Physics Today* **63**, 27–32 (2010).
13. Bohm, D. A suggested interpretation of the quantum theory in terms of "hidden" variables. I. *Physical Review* **85**, 166 (1952).
14. Bohm, D. *Quantum Theory* (Courier Corporation, 1989).
15. Wyatt, R. E. *Quantum Dynamics with Trajectories: Introduction to Quantum Hydrodynamics*, vol. 28 (Springer Science & Business Media, 2005).
16. Sanz, Á. S. & Miret-Artés, S. *A Trajectory Description of Quantum Processes. I and II. Applications: A Bohmian Perspective* (Springer, 2013).
17. Takahashi, S. & Takatsuka, K. Phase quantization of chaos in the semiclassical regime. *The Journal of Chemical Physics* **127** (2007).
18. Yamashita, T. & Takatsuka, K. Phase quantization of chaos and the role of the semiclassical amplitude factor. *Progress of Theoretical Physics Supplement* **166**, 56–69 (2007).
19. Takatsuka, K., Takahashi, S., Koh, Y. W. & Yamashita, T. Energy quantization of chaos with the semiclassical phases alone. *The Journal of Chemical Physics* **126** (2007).
20. Takatsuka, K. & Takahashi, S. Towards many-dimensional real-time quantum theory for heavy-particle dynamics. II. Beyond semiclassics by quantum smoothing of the singularity in quantum-classical correspondence. *Physical Review A* **89**, 012109 (2014).

Real-Valued Path Integral as Green's Functions of the Schrödinger Equation

6

Abstract

The relationship between the summation over possible events and distribution functions is examined through the path-integral formalism. The path integral is known to provide an explicit representation of the "sum over events" or "sum over histories" contained in the distribution amplitude function. After reviewing the Feynman path integral and its mathematical difficulty, we show:

1. A real-valued path integral is formulated as the Green function of the real-valued Schrödinger equation for the Schrödinger vector.
2. Unlike the complex Feynman path integral, the real path integral possesses a (skewed) Wiener measure.
3. Besides the meaning of the Schrödinger dynamics as a sum over paths, it highlights the presence of stochasticity underlying quantum mechanics and the strong similarity with the Feynman–Kac formula in stochastic process theory, thereby strongly suggesting stochastic dynamics in nature behind quantum mechanics.

Keywords

Feynman path integral · Real-valued path integrals · Feynman-Kac formula · Wiener measure · Stochasticity in quantum mechanics

In this chapter, the relationship between the summation over possible events and distribution functions is examined through the path integral formalism. The path integral provides an explicit representation of the "sum over events" or "sum over histories" contained in the distribution amplitude function. This is analyzed in

K. Takatsuka, *Structure of the Schrödinger Dynamics*,
https://doi.org/10.1007/978-981-92-1910-0_6

terms of real-valued path integral serving as Green's function of the real-valued Schrödinger equation.

6.1 Feynman Path Integral

We begin with an overview of quantum path integral—namely, Feynman path integral—from the perspective of summation over events [1–4]. In a sense, this represents the underlying structure behind the Schrödinger wavefunction. Richard Feynman is often quoted as saying that "nobody understands quantum mechanics." Nevertheless, his path-integral formulation is both eloquent and highly suggestive.

At first glance, the Feynman path integral appears to be a Riemann sum over successively added sample paths. In reality, however, it does not possess the integral measure and thereby absolute convergence is not assured. Compared with the Feynman-Kac formula for diffusion processes [5] (Eq. (4.3.9)), this represents a fundamental mathematical difference. Historically, the Feynman path integral inspired the development of the Feynman-Kac formula, yet the latter possesses desirable convergence properties absent in the former. Since the Feynman path integral is conceived as Green's functions of the Schrödinger equation, it is reasonable to expect that some form of convergence property can be found.

6.1.1 The Feynman Kernel

Let us review the basic aspects of the Feynman path integral [1–4]. The time evolution operator in quantum mechanics is

$$\hat{U}(t, t_0) = \exp\left(\frac{1}{i\hbar}\hat{H}(t - t_0)\right), \tag{6.1.1}$$

and its q representation

$$\begin{aligned} K(q_2, t_2; q_1, t_1) &= \langle q_2| \hat{U}(t_2, t_1) |q_1\rangle \\ &= \langle q_2| \exp\left(\frac{1}{i\hbar}\hat{H}(t_2 - t_1)\right) |q_1\rangle \end{aligned} \tag{6.1.2}$$

is what we need here. Due to the limit

$$\lim_{t_2 \to t_1} K(q_2, t_2; q_1, t_1) \to \delta(q_2 - q_1), \tag{6.1.3}$$

$K(q_2, t_2; q_1, t_1)$ is the Green's function of the Schrödinger equation such that

$$\left(i\hbar\frac{\partial}{\partial t} - \hat{H}\right) K(q_2 t_2; q_1 t_1) = i\hbar\delta(q_2 - q_1)\delta(t_2 - t_1). \tag{6.1.4}$$

This Green's function is called the Feynman kernel, and if $K(q,t;q_0,t_0)$ is known, the time evolution of any (complex) Schrödinger function can be automatically obtained by integrating over the kernel as

$$\psi(q,t) = \int dq_0 K(q,t;q_0,t_0)\psi(q_0,t_0). \tag{6.1.5}$$

Similarly to Eq. (4.3.8), for an infinitesimal time interval Δt, the Schrödinger equation can be formally integrated in terms of a path integral over Δt as

$$\psi(x,t+\Delta t) = \left(\frac{m}{2\pi i\hbar\Delta t}\right)^{1/2} \times \int_{-\infty}^{\infty} dy \exp\left[\frac{i}{\hbar}\left(\frac{m}{2}(x-y)^2/\Delta t - V(y,t)\Delta t\right)\right]\psi(y,t). \tag{6.1.6}$$

Then, returning to the expression of Eq. (6.1.2), we have

$$\langle q_2|\exp\left(\frac{1}{i\hbar}\hat{H}\Delta t\right)|q_1\rangle = \left(\frac{m}{2\pi i\hbar\Delta t}\right)^{1/2} \exp\left[\frac{i}{\hbar}\left(\frac{m}{2}(q_2-q_1)^2/\Delta t - V(q_1,t)\Delta t\right)\right] \tag{6.1.7}$$

Thus, with $\Delta t = \lim_{t_2\to t_1}(t_2-t_1)$, by using this twice, we obtain

$$\int dq_2 \left\langle q_3 \left| \exp\left(\frac{1}{i\hbar}\hat{H}\Delta t\right)\right| q_2\right\rangle\left\langle q_2 \left|\exp\left(\frac{1}{i\hbar}\hat{H}\Delta t\right)\right| q_1\right\rangle = \left\langle q_3 \left|\exp\left(\frac{1}{i\hbar}\hat{H}\,(2\Delta t)\right)\right| q_1\right\rangle. \tag{6.1.8}$$

Therefore, by repeatedly using this short-time insertion formula, we can construct the kernel for a finite time interval. Formally, this can be written as

$$K(q,t:0,0) = \lim_{N\to\infty}\int_{-\infty}^{\infty} dq_1\cdots dq_{N-1}\left(\frac{m}{2\pi i\hbar\Delta t}\right)^{(N-1)/2} \times \exp\left[\sum_{k=1}^{N-1}\frac{i}{\hbar}\left(\frac{m}{2}\frac{(q_{k+1}-q_k)^2}{\Delta t} - V(q_k,t_k)\Delta t\right)\right], \tag{6.1.9}$$

where $dq_k = q_{k+1} - q_k$, $q_N = q$, $q_0 = 0$, and $\Delta t = t/N$.

6.1.2 From Successive Integrals to a Path Integral: The Sum Over Possible Events

Up to this point, an issue is whether $\int_{-\infty}^{\infty} dq_k$ can be properly evaluated by taking Δt to be sufficiently small. However, this is only the matter of approximation and of no theoretical problem. In Eq. (6.1.9), a sequence such as $q_0 \to q_1 \to q_2 \to q_3 \to \cdots \to q_N$ appears, which, in the limit $\Delta t \to 0$, looks like a smooth curve, and there are infinitely many such broken lines. Feynman's idea was to reinterpret the Riemann sum over $\prod_k \int_{-\infty}^{\infty} dq_k$ as a Riemann sum over the paths $q_0 \to q_1 \to q_2 \to q_3 \to \cdots \to q_N$, and to consider the entire kernel $K(q, t : 0, 0)$ as a sum over all those possible paths from the initial point $(0, 0)$ to the final point (q, t) [1,2]. This brilliant idea is, in essence, the same as Wiener's path integral in spirit. The part

$$\frac{m}{2}\left(\frac{q_{k+1} - q_k}{\Delta t}\right)^2 - V(q_k, t_k) \tag{6.1.10}$$

of the exponent in Eq. (6.1.9) corresponds to the classical action integral in the limit $\Delta t \to 0$ of, and thus, in order to implicitly express the idea of "integrating over the path," it is customary and formally to write

$$K(q, t : 0, 0) = \int_{\Omega[q,t:0,0]} D\mathbf{q}(\tau) \exp\left[\frac{i}{\hbar} S(\mathbf{q}(\tau))\right]. \tag{6.1.11}$$

In this notation, each path is represented by $\mathbf{q}(\tau)$, meaning that at time τ, the path passes through $\mathbf{q}(\tau)$. The exponent

$$S(\mathbf{q}(\tau)) = \int_0^t \left[\frac{1}{2} m \left(\frac{d\mathbf{q}(\tau)}{d\tau}\right)^2 - V(\mathbf{q}(\tau))\right] \tag{6.1.12}$$

represents the form of the action integral along the path $\mathbf{q}(\tau)$. In Eq. (6.1.11), the infinitesimal volume elements necessary for the Riemann sum are written as

$$D\mathbf{q}(\tau) \tag{6.1.13}$$

which is complex-valued. It remains a mathematical problem how to formulate this quantity rigorously. otherwise, the general solution to the Schrödinger equation can be obtained in just a "formal manner" as an explicit expression of a Green function.

The method of interpreting $\prod_k \int_{-\infty}^{\infty} dq_k$ in Eq. (6.1.9) as a "sum over possible paths" was originally developed by Norbert Wiener in the study of diffusion processes, as explained below Eq. (4.3.5). It is often noted that Feynman was unaware of this work. After the proposal of Feynman's path integral, Wiener's formulation was extended by Mark Kac to diffusion processes in the presence of a potential function

$V(q)$, leading to the Feynman-Kac formula (Eq. (4.3.10)). However, problems that do not arise in the real-valued path integral of the Feynman-Kac formula do appear in Feynman's complex-valued path sum. In the Feynman-Kac formulation, each path is assigned a weight given by the Wiener measure, Eq. (4.3.11). By contrast, the Feynman path integral lacks an integration measure in the usual sense—such as the Riemann or Stieltjes measure—leaving open the question of whether it should properly be regarded as an integral.

6.1.3 Oscillatory Integrals: Riemann-Lebesgue Lemma and Stationary Phase

Since $K(q, t : 0, 0)$ of Eq. (6.1.11) contains the Planck constant $\hbar$, which is very small, the integrand exhibits rapid oscillations both along a given integration path and under small deformations of that path. Such integrals fall into the category of oscillatory integrals, whose mathematical treatment is provided by asymptotic analysis [6–8].

One of the most well-known results of asymptotic analysis is the method of stationary phase:

$$\begin{aligned}&\int g(x)\exp\left(ikf(x)\right)\\&=\sum_{x_0} g(x_0)\exp\left[ikf(x_0)+\mathrm{sign}(f''(x_0))\times i\pi/4\right]\left(\frac{2\pi}{k\,|f''(x_0)|}\right)\\&+o\left(k^{-1/2}\right),\end{aligned} \tag{6.1.14}$$

where x_0 are the (possibly multiple) stationary points satisfying

$$f'(x_0)=0 \tag{6.1.15}$$

for the phase function f. Applying this result to Eq. (6.1.11), one finds that only those paths satisfying the stationary condition

$$\delta S(\mathbf{q}(\tau)=0 \tag{6.1.16}$$

can make a significant contribution to the integral, and thereby, they are to be selected. (Incidentally, the Schrödinger functions were found to satisfy the stationary condition of Eq. (5.1.40).) The stationary solutions of Eq. (6.1.16) correspond precisely to the classical trajectories determined by the principle of least action. Geometrically, this means that in the vicinity of a classical path, the rapid oscillations of the integral are suppressed, and the action $S(\mathbf{q}(\tau)$ varies smoothly.

If no stationary phase point exists within the integration domain, the integral asymptotically vanishes according to the Riemann-Lebesgue lemma

$$\int g(x) \exp(-izx)\, dx \to 0 \qquad \text{as } |z| \to \infty, \tag{6.1.17}$$

which, in the context of Eq. (6.1.11), implies that $K(q, t : 0, 0) \to 0$ ($o(\hbar^{1/2})$) in the limit $\hbar \to 0$, unless classical path is involved. Contributions from nonclassical paths cancel out due to the rapid oscillations of the phase.

Besides, in Eq. (6.1.14), it may occur not only that $f'(x_0) = 0$, but also that $f''(x_0) = 0$. This situation arises when two or more stationary points coalesce. In such cases, one must unfold the underlying geometric structure around the stationary point and evaluate the integral in terms of higher-order derivatives of $f(x)$ at $x = x_0$. This procedure is commonly referred to as a uniform approximation (see, e.g., [9–13]). Although we do not pursue these advanced approximation methods here, they are collectively known as semiclassical mechanics. In this way, the path integral formulation of Feynman provides a profound bridge between quantum mechanics and classical mechanics.

6.1.4 Nonexistence of Integral Measure and No Absolute Convergence

Returning to the Feynman path integral, Eq. (6.1.11), all possible paths are assigned an "imaginary" weight $D\mathbf{q}(\tau)$. However, with this weight being complex, the contributions from individual paths undergo highly oscillatory behavior, requiring the inclusion of an increasingly large (indeed, infinite) number of paths to achieve mutual cancellation. As a result, absolute convergence in the sense of ordinary Riemann integration cannot be expected. Nevertheless, this mathematical deficiency does not diminish the gigantic physical significance of the Feynman path integral.

The origin of this mathematical difficulty lies in the absence of a proper measure required to define the path sum rigorously. Even if the sum appears to converge, paths may still produce oscillatory contributions nearby, and even those paths that violate relativistic constraints are not excluded. This is the essential difference from the Feynman-Kac path integral.

It is worth noting that overcoming the lack of an integration measure in the Feynman path integral remains a challenging problem in mathematical physics [4, 14, 15].

6.1.5 Complex Time and Quantum Partition Function

If we try to suppress the oscillatory path integral in the time evolution operator

$$\hat{U}(t) = \exp\left(-\frac{it}{\hbar}\hat{H}\right), \tag{6.1.18}$$

we simply transform from real time to imaginary as

$$\frac{it}{\hbar} \to \beta = \frac{1}{k_B T} \tag{6.1.19}$$

by coordinate rotation. $\hat{U}(t)$ is then

$$\hat{U}(\beta) = \exp\left(-\beta\hat{H}\right) \tag{6.1.20}$$

with no oscillatory behavior in principle. Thus, as above, by taking T as temperature and k_B as the Boltzmann constant, it becomes possible to connect to statistical mechanics, since the trace of the "evolution operator"

$$\begin{aligned} Z(\beta) &= \int dq \left\langle q \left| \exp\left(-\beta\hat{H}\right) \right| q \right\rangle \\ &= Tr \exp\left(-\beta\hat{H}\right) \end{aligned} \tag{6.1.21}$$

looks a partition function in quantum representation. Indeed, various approaches have been proposed to carry out this integral as a powerful method for quantum many-body problems. Many works are found mainly in the literature of chemical physics of condensed phase.

What about the inverse problem? We start from the calculation of $\hat{U}(\beta) = \exp\left(-\beta\hat{H}\right)$ avoiding the oscillatory integral and bring β back to the time t by analytic continuation. However, there is no guarantee that the rotation of the β-axis can always be performed correctly, since the relevant analyticity, depending on the object, is generally not known.

6.2 Real-Valued Quantum Path Integral

As seen above, due to the existence of imaginary numbers, the Feynman path integral has both great advantages and certain demerits. To eliminate those demerits as much as possible, let us return to the real-valued Schrödinger equation and consider the path integral as a Green function of the real differential equation.

6.2.1 Schrödinger Equation as Coupled Diffusion Equations

By multiplying $-\mathbf{J}$ from the left to the real Schrödinger equation (5.1.44), we rewrite it as

$$\frac{\partial}{\partial t}\begin{pmatrix} \phi_r(q,t) \\ \phi_c(q,t) \end{pmatrix} = \left(\frac{\hbar}{2m}\nabla^2 - \frac{V(q)}{\hbar}\right)\mathbf{J}\begin{pmatrix} \phi_r(q,t) \\ \phi_c(q,t) \end{pmatrix} \tag{6.2.1}$$

or more concisely as

$$\frac{\partial}{\partial t}\bar{\psi}(q,t) = \left(\frac{\hbar}{2m}\nabla^2 - \frac{1}{\hbar}V(q)\right)\mathbf{J}\bar{\psi}(q,t). \tag{6.2.2}$$

This expression is quite akin to the diffusion equation (4.3.7) with correspondence $D = \hbar/2m$ and $\lambda = 1/\hbar$ (see also (5.2.33) for reference). However, we see that the symplectic unit vector $\mathbf{J}$ exists in Eq. (6.2.2), and the only difference between the two equations is due to $\mathbf{J}$. Or the (mathematical) leap to quantum mechanics is yielded by this $\mathbf{J}$.

Equation (6.2.2) can be written in the form:

$$\bar{\psi}(t+\Delta t) = \exp\left[\left(\frac{\hbar}{2m}\nabla^2 - \frac{1}{\hbar}V(q)\right)(\mathbf{J}\Delta t)\right]\bar{\psi} \tag{6.2.3}$$

for a short time interval Δt, and Eq. (4.3.8) modified only with $\mathbf{J}$ can be written in the form:

$$\bar{\psi}(x,t+\Delta t) = \left(\frac{m}{2\pi\hbar\mathbf{J}\Delta t}\right)^{1/2} \times \int_{-\infty}^{\infty} dy \exp\left[\frac{\mathbf{J}}{\hbar}\left(\frac{m}{2}\frac{(x-y)^2}{\Delta t} - V(y,t)\Delta t\right)\right]\bar{\psi}(y,t) \tag{6.2.4}$$

or

$$\bar{\psi}(x,t+\Delta t) = \left(\frac{m}{2\pi\hbar\mathbf{J}\Delta t}\right)^{1/2} \times \int_{-\infty}^{\infty} dy \exp\left[\frac{\mathbf{J}^T}{\hbar}\left(-\frac{m}{2}\frac{(x-y)^2}{\Delta t} + V(y,t)\Delta t\right)\right]\bar{\psi}(y,t). \tag{6.2.5}$$

Here, the properties of (5.1.47) and (5.1.48) have been used.

Notice that Eq. (6.2.4) is simply the equation obtained by replacing i with $\mathbf{J}$ in (6.1.6). In other words, it can be developed in parallel with the Feynman-Kac formulas (4.3.9), (4.3.10), and (4.3.11), and therefore, the Green function of the real-valued Schrödinger equation (6.2.1) can be obtained as

$$\mathbf{G}(q,t:0,0) = \lim_{N\to\infty}\int_{-\infty}^{\infty}\Delta q_1\cdots\Delta q_N\left(\frac{m}{2\pi\hbar\mathbf{J}\Delta t}\right)^{N/2} \times \exp\left[\sum_{k=1}^{N}\frac{1}{\hbar}\mathbf{J}^T\left(-\frac{m}{2}\frac{(\Delta q_k)^2}{\Delta t} + V(q_k,t_k)\Delta t\right)\right], \tag{6.2.6}$$

where $\Delta q_k = q_k - q_{k-1}$, $q_N = q$, $q_0 = 0$. Thus, here, in the form of a formal path integral

$$\mathbf{G}(q, t : 0, 0) = \int_{\Omega[q,t:0,0]} \exp\left[\frac{\mathbf{J}^T}{\hbar}\int_0^t V(s, X(s, \omega))\, ds\right] \times d\mathbf{P}_{W[q,t:0,0]}(\omega) \tag{6.2.7}$$

can be expressed as follows. Here, in this path integral, there is a real-valued formal measure:

$$d\mathbf{P}_{W[q,t:0,0]}(\omega) = \lim_{N\to\infty}\left(\frac{m}{2\pi\hbar\mathbf{J}\Delta s}\right)^{N/2} \times \exp\left[\sum_{k=1}^{N}\Delta s\left(-\frac{m}{2\hbar}\left(\frac{\Delta q_k}{\Delta s}\right)^2\right)\mathbf{J}^T\right]\prod_{k=1}^{N} dq_k. \tag{6.2.8}$$

We will call $d\mathbf{P}_{W[q,t:0,0]}(\omega)$ the skew Wiener measure.

6.2.2 Skew Wiener Measure and Its Properties

Here, let us examine in some detail the properties of the distorted Wiener measure by comparing it with the Wiener measure in the Feynman-Kac formula, Eq. (4.3.11).

First, in both $d\mathbf{P}_{W[q,t:0,0]}(\omega)$ in Eq. (6.2.8) and $dP_{W(q,t:0,0)}(\omega)$ in Eq. (4.3.11), it should be noted that, in order for these expressions to serve as measures in the infinitesimal limit, the increments Δq_k and the time interval Δs (or Δt) cannot be taken to the infinitesimal limit independently in an arbitrary manner. As can be seen from the expressions, it is necessary to take the limit in a correlated way such that

$$\frac{(\Delta q_k)^2}{\Delta t} \tag{6.2.9}$$

has a finite average value. This fact is well known in the theory of Brownian motion, but it also carries important implications in quantum theory. The physical consequences of this will be discussed in detail in Chap. 9.

Next, let us examine the role of the exponential factors appearing in $d\mathbf{P}_{W[q,t:0,0]}(\omega)$ and $dP_{W(q,t:0,0)}(\omega)$. In the Feynman-Kac formula, the magnitude of $dP_{W(q,t:0,0)}(\omega)$ depends on the choice of $(\Delta q_k)^2/\Delta t$ and undergoes large fluctuations through the exponential factor. The sampling paths in $dP_{W(q,t:0,0)}(\omega)$ should be chosen such that $(\Delta q_k)^2/\Delta t$ takes values of the order of the diffusion constant D. However, even if it is chosen to be large, the exponential factor automatically suppresses such contributions, resulting in a small measure. This implies that, due to the friction acting between the Brownian particle and the medium, the probability that the particle travels a long distance within a short time is small, even in the absence of a potential, and such paths are effectively excluded at the level of the measure.

On the other hand, in the skewed Wiener measure $d\mathbf{P}_{W[q,t:0,0]}(\omega)$, the exponential factor is a 2×2 matrix. In order to estimate its magnitude, we compute its determinant such that

$$\begin{aligned} &\det \exp\left[\sum_{k=1}^{N} \Delta s \left(-\frac{m}{2\hbar}\left(\frac{\Delta q_k}{\Delta s}\right)^2\right)\mathbf{J}^T\right] \\ &= \exp\left(Tr\left[\sum_{k=1}^{N} \Delta s \left(-\frac{m}{2\hbar}\left(\frac{\Delta q_k}{\Delta s}\right)^2\right)\mathbf{J}^T\right]\right) \\ &= 1. \end{aligned} \tag{6.2.10}$$

This result follows from $Tr\mathbf{J} = 0$. In this way, regardless of the path chosen, the contribution from the exponential factor yields a constant value in this sense, making it closer in character to a Riemann measure in simple integrals over Euclidean space. This reflects the fact that, in the case of quantum path integrals, the Schrödinger equation is constructed in a simple Euclidean space, without any consideration of the Lorentz invariance or space-time curvature. Consequently, notions such as a velocity limit up to the speed of light or "friction with the vacuum" are not included in the concept of Schrödinger dynamics from the outset.

However, as emphasized repeatedly, at the stage of applying Eq. (6.2.4), it is necessary that $(\Delta q_k)^2/\Delta s$ be chosen so that its average magnitude is of the order of $\hbar/m$, or

$$\left\langle \frac{1}{2}m\left(\frac{\Delta q_k}{\Delta s}\right)^2\right\rangle \sim \frac{1}{2}\frac{\hbar}{\Delta s}, \tag{6.2.11}$$

which imposes a condition on the kinetic energy distribution of sampled paths to be chosen in the path integrations.

6.2.3 Classical Trajectories as Steepest Descent Values

One of the most important conclusions of the Feynman path integral is that, among an infinite number of sampled paths, those satisfying the stationary phase condition are naturally selected as classical trajectories (Eq. (6.1.16)). In the real-valued path integral, in Eq. (6.2.6), the exponential factor

$$\lim_{N\to\infty} \exp\left[-\sum_{k=1}^{N-1} \frac{1}{\hbar}\mathbf{J}^T\left(\frac{m}{2}v(\omega)^2 - V(q_k, t_k)\right)\Delta t\right]$$

$$= \exp\left[-\frac{1}{\hbar}\mathbf{J}^T \int \left(\frac{m}{2}v(\omega)^2 - V(q,s)\right) ds\right] \tag{6.2.12}$$

leads, under the limit $\hbar \to 0$, to the condition for the method of steepest descent. Namely, one selects trajectories satisfying

$$\delta\left(\frac{m}{2}v(\omega)^2 - V(q,s)\right) = 0 \tag{6.2.13}$$

This is identical to the stationary phase condition given in Eq. (6.1.16).

6.3 Perturbative Approximation from the Path Integral

The path integral formulation possesses a characteristic approximation and computational methods that cannot be considered at the level of the Schrödinger equation. As one useful example of approximation theory, we introduce perturbation theory.

First, returning to Eq. (6.2.6), we rewrite it as

$$\mathbf{G}(q,t:0,0) = \lim_{N\to\infty}\int_{-\infty}^{\infty} \Delta q_1 \cdots \Delta q_{N-1}\left(\frac{m}{2\pi\hbar\mathbf{J}\Delta t}\right)^{(N-1)/2}$$

$$\times \exp\left[\sum_{k=1}^{N-1}\frac{1}{\hbar}\mathbf{J}^T\left(-\frac{m}{2}\frac{(\Delta q_k)^2}{\Delta t} + V_0(q_k,t_k)\Delta t + \Delta V(q_k,t_k)\Delta t\right)\right] \tag{6.3.1}$$

where

$$\Delta V(q_k,t_k) = V(q_k,t_k) - V_0(q_k,t_k). \tag{6.3.2}$$

Here, $V_0(q_k,t_k)$ is taken as a zeroth-order reference potential that is supposed to be solved relatively easily, while $\Delta V(q_k,t_k)$ is treated as a perturbing potential.

In this case, the Green function in path integral form becomes

$$\mathbf{G}(q,t:0,0) = \int_{\Omega[q,t:0,0]} \exp\left[\frac{\mathbf{J}^T}{\hbar}\int_0^t \Delta V(s, X(s,\omega))\, ds\right]$$

$$\times\, d\mathbf{P}_{W[q,t:0,0]}(\omega) \tag{6.3.3}$$

and the Wiener measure is rewritten with the weight of the zeroth-order potential as

$$d\mathbf{P}_{W[q,t:0,0]}(\omega) = \lim_{N\to\infty}\left(\frac{m}{2\pi\hbar\mathbf{J}\Delta s}\right)^{(N-1)/2}$$

$$\times \exp\left[\sum_{k=1}^{N}\Delta s\left(-\frac{m}{2\hbar}\left(\frac{\Delta q_k}{\Delta s}\right)^2 + V_0(q_k,t_k)\right)\mathbf{J}^T\right]\prod_{k=1}^{N-1} dq_k. \tag{6.3.4}$$

This simple rewriting becomes useful when the dynamics under $V_0(q_k, t_k)$ are known, and the paths are slightly distorted by a small perturbation ΔV.

Likewise, the total system under consideration can be divided into two interacting subsystems (like the principal system and the environment), and the effect of the environment can be incorporated into the path integral of the principal system. This is known as the Feynman-Vernon influence functional method [16] and is a widely used technique in path integral approaches. We refer the reader to other texts for details [2–4].

References

1. Feynman, R. P. Space-time approach to non-relativistic quantum mechanics. *Reviews of Modern physics* **20**, 367 (1948).
2. Feynman, R. P., Hibbs, A. R. & Styer, D. F. *Quantum Mechanics and Path integrals* (Courier Corporation, 2010).
3. Schulman, L. S. *Techniques and Applications of Path integration* (Courier Corporation, 2012).
4. Kleinert, H. *Path Integrals in Quantum Mechanics, Statistics, Polymer Physics, and Financial Markets* (World Scientific Publishing Company, 2006).
5. Kac, M. On distributions of certain wiener functionals. *Transactions of the American Mathematical Society* **65**, 1–13 (1949).
6. Erdélyi, A. *Asymptotic Expansions*. 3 (Courier Corporation, 1956).
7. Bleistein, N. & Handelsman, R. A. *Asymptotic Expansions of Integrals* (Ardent Media, 1975).
8. Murray, J. D. *Asymptotic Analysis*, vol. 48 (Springer Science & Business Media, 2012).
9. Levit, S. & Smilansky, U. The hamiltonian path integrals and the uniform semiclassical approximations for the propagator. *Annals of Physics* **108**, 165–197 (1977).
10. Klauder, J. R. Global, Uniform, semiclassical approximation to wave equations. *Physical Review Letters* **56**, 897 (1986).
11. Vaníček, J. & Heller, E. J. Uniform semiclassical wave function for coherent two-dimensional electron flow. *Physical Review E* **67**, 016211 (2003).
12. Brack, M. & Bhaduri, R. *Semiclassical Physics* (CRC Press, 2018).
13. Child, M. S. *Semiclassical Mechanics with Molecular Applications* (Claredon Press, Oxford, 1991).
14. Klauder, J. R. & Daubechies, I. Quantum mechanical path integrals with Wiener measures for all polynomial Hamiltonians. *Physical Review Letters* **52**, 1161 (1984).
15. Daubechies, I. & Klauder, J. R. Quantum-mechanical path integrals with wiener measure for all polynomial Hamiltonians. II. *Journal of Mathematical Physics* **26**, 2239–2256 (1985).
16. Feynman, R. P. & Vernon Jr, F. L. The theory of a general quantum system interacting with a linear dissipative system. *Annals of Physics* **281**, 547–607 (2000).

Summation over Events and Distribution Functions

7

Abstract

Although the solution of the real-valued Schrödinger equation is a vector (the Schrödinger vector), and the corresponding Green function is a matrix containing **J**, the symplectic standard matrix, and the paths that transport this vector are scalar stochastic trajectories in configuration space. Based on these properties, we discuss the quantum superposition principle as a characteristic of quantum distribution amplitude function, including the Schrödinger's cat paradox.

Keywords

Superposition principle · Hierarchy of distribution functions · Reduced density matrix · Schrödinger's cat

The fundamental physical picture obtained from the real-valued quantum path integral representations (Eq. (6.2.6) or Eq. (6.2.7)) in the previous chapter is as follows. Although the solution to the real Schrödinger equation is a vector (the Schrödinger vector) and the corresponding Green function is a matrix containing **J** as an element, the paths that transport this vector are scalar stochastic trajectories in configuration space. This point was not self-evident to the author, since he was afraid of the possibility that the paths themselves might be vector-valued naturally arises.

Each path that drives the Green function corresponds to an individual "physical event." Whether or not these paths obey a specific physical law, such as Newtonian mechanics, is a separate issue. Feynman emphasized that the essence of quantum theory lies in coherent summation over all paths in a democratic manner. Summing over all paths is equivalent to summing over all "possible events." This principle also holds in the classical probabilistic framework through the Feynman-Kac formula. In diffusion processes, even when the exact shape of the path is unknown, one

K. Takatsuka, *Structure of the Schrödinger Dynamics*,
https://doi.org/10.1007/978-981-92-1910-0_7

can predict the spatiotemporal evolution of the distribution function by summing over all paths, each weighted by its associated statistical weight and integration measure. This is precisely the claim of the Wiener path integral and the Feynman-Kac formula. However, in this classical stage of hierarchy, there are no interference terms; there is only summation. In other words, the distribution function of Brownian motion, ρ_d, is obtained as a sum of all possible Brownian paths, which is not the case for quantum dynamics. To deepen our understanding of the essence of quantum theory, let us now examine the role and structure of this "summation" underlying distribution functions.

7.1 Structural Similarity Between the Diffusion Equation and the Schrödinger Equation

By this stage, the similarities and differences between classical diffusion phenomena and Schrödinger dynamics have become sufficiently clear, which are summarized below.

On the Distribution Function

1. In classical diffusion, the description is completed with the distribution ρ_d that satisfies the diffusion equation [Eq. (4.2.3)].
2. In Schrödinger dynamics, one first assumes the existence of a classically observable probability density function ρ [Eq. (5.1.1)] and then proceeds to a deeper level by introducing the distribution amplitude function $\bar{\psi}$, defined as the square-root vector of ρ [Eqs. (5.1.2), (5.1.44), (6.2.2)].
3. The function $\bar{\psi}$ is essentially identical to the complex Schrödinger wavefunction ψ [Eqs. (5.1.52) and (5.1.53)].

On the Path Integrals

1. In classical diffusion, the Green function G_D appearing in the Feynman-Kac formula governs the time evolution of ρ_d [Eqs. (4.3.9), (4.3.10)]. The function G_D transports scalar values of ρ_d defined on points in configuration space.
2. In Schrödinger dynamics, the real-valued quantum Green function G drives the time evolution not of ρ, but of the real-valued Schrödinger vector $\bar{\psi}$ [Eqs. (6.2.6) and (6.2.7)]. The real-valued Schrödinger equation has a form strikingly similar to that of the classical diffusion equation. Yet G transports vector-valued quantities $\bar{\psi}$ in configuration space.

Thus, even for the Schrödinger vector, the following statement holds: Regardless of the specific form of the paths, if all possible paths are summed with their associated statistical weights and integration measures, the correct spatiotemporal evolution can be predicted. Consequently, one finds that classical diffusion $\rho_d(q, t)$ and quantum theory $\bar{\psi}(q, t)$ share mathematically analogous structures, albeit at different hierarchical levels.

7.2 On the Principle of Superposition in Distribution Functions

7.2.1 Distribution Function Used for Prediction and the Distribution Obtained from Surveying a Sample Population

Let us recall that when we casually refer to a "statistical distribution function," there are in fact two distinct types to consider. One type is exemplified by the case in which we take, as a statistical ensemble, all the students in a certain grade at an elementary school, "survey" the distribution of their heights, and represent it by a function like the Gaussian or a similar form. In contrast, the distributions employed in the context of quantum mechanics take as their frequency a set of various "possible" events (for example, represented by paths). These events, although independent in the sense that they occur in separate trials under the same environment, are nevertheless all summed and incorporated into a single function. The reader may recall the function $\Gamma(q, p, t)$ in the Liouville equation, as well as the discussion of distributions in the paragraph immediately following Eq. (3.2.35). The primary objective here is not the quantitative description of surveyed outcomes, but rather mathematical prediction. Both the quantum-mechanical path integrals and the Feynman-Kac formula for diffusion fall into this category.

Even under such circumstances, in the case of classical events, the phenomena obtained by "surveying" what actually occurs approach those predicted by the well designed "mathematical" distribution function as the frequency of observation increases. Likewise, for pure mathematical distributions, it is to be expected that more accurate predictions can be achieved as the set of assumed events is refined and the number of samples increases. In this sense, a principle of superposition (or summation) also exists for classical distribution phenomena. In quantum mechanics, on the other hand, the principle of superposition holds at the level of the quantum-mechanical path integral and of $\bar{\psi}(q, t)$ but does not hold at the level of $\rho(q, t)$, which is accessible to classical observation. We shall verify these points in what follows.

7.2.2 Hierarchy of the Level of Superposition

Classical Diffusion

In the diffusion equation, the distribution function $\rho_d(q, t)$ is proportional to the frequency of finding a Brownian particle at point q at time t (more precisely, elapsed time counted from the initial moment). Suppose that multiple gates are introduced in the diffusion process (denoted by $A, B, C, \cdots$, through which a particle can pass only once exclusively). Then, the distribution of particles reaching (q, t) via each gate satisfies $\rho_d(q, t) = \rho_A(q, t) + \rho_B(q, t) + \cdots$, so that a simple additive relation holds. This basic fact is guaranteed by the linearity of the diffusion equation.

It is important to note again that, in classical diffusion, the quantity $\rho_d(q,t)$ obtained in this way directly corresponds to an observable quantity.

Schrödinger Dynamics

In analogy with the classical diffusion case, let us introduce the similar gates along the possible paths of a particle. If an initial distribution amplitude $\bar{\psi}(q,0)$ spreads over time t, then all possible contributions corresponding to passage through the respective gates must be summed. Accordingly, one should write $\bar{\psi}(q,t) = \bar{\psi}_A(q,t) + \bar{\psi}_B(q,t) + \cdots$, where the sum is a vector sum.

A crucial point to note here again is that t represents elapsed time from the initial condition, not a fixed instant. The quantities $\bar{\psi}_A(q,t)$, $\bar{\psi}_B(q,t), \ldots$ do not indicate that the particle has arrived at point q at the same time t via different gates. Rather, they represent different possible events—"this may have occurred" or "that may have occurred"—and only by summing over all such possibilities does one obtain a quantity related to the overall frequency. Thus, $\bar{\psi}(q,t)$ does not represent an individual event, but rather the sum of all possible events.

Such a summation process is necessarily involved at some stage in constructing any distribution function. The essential question is at which hierarchical level does this summation take place. At the stage of summation, each statistical sample (e.g., an electron, atom, or molecule) used to construct the distribution is not supposed to pass through plural gates at a time and undergo self-interference as in the double-slit experiment.

The Schrödinger vector $\bar{\psi}(q,t)$ is a two-component real vector, which may, for example, be represented in terms of magnitude and angle. At the level of Schrödinger dynamics, $\bar{\psi}(q,t)$ propagates and accumulates as a vector spreading at each point (q,t). However, if one wishes to obtain classically observable quantities, one must return to the classical distribution function via $\rho(q,t) = \bar{\psi}(q,t)^T \bar{\psi}(q,t)$, which is a scalar quantity. As such, in classical diffusion dynamics, all information appears and is observed at the level of $\rho_d(q,t)$. In contrast, quantum theory requires descending to the level of the Schrödinger vector in order to trace the causal structure of the dynamics and then taking the sum of the resulting contributions. Why quantum dynamics requires this deeper level is not fully understood; one can only state that nature appears to be structured in this way.

Moreover, in the decomposition $\rho(q,t) = \bar{\psi}(q,t)^T \bar{\psi}(q,t)$, there is no logical necessity that $\bar{\psi}(q,t)$ be a two-component vector; in nonrelativistic quantum mechanics, this happens to suffice. In principle, deeper hierarchical structures are mathematically and physically possible, as in relativistic quantum theory.

On the Fringe Pattern in the Double-Slit Experiment

From the relation $\bar{\psi}(q,t) = \bar{\psi}_A(q,t) + \bar{\psi}_B(q,t) + \cdots$ together with $\rho(q,t) = \bar{\psi}(q,t)^T \bar{\psi}(q,t)$, it is natural, from a mathematical standpoint, that interference fringes appear at the level of the classical hierarchical stage of distribution density function by stepping from $\bar{\psi}(q,t)$ up to $\rho(q,t)$.

Accordingly, the appearance of interference fringes in the "accumulated" distribution resulting from the emission of individual electrons in the double-slit experiment is to be naturally expected, in that a distribution function represents the entirety of possible individual events. The point to be stressed again is as follows: In Brownian motion, the summing-up principle is realized at the level of $\rho_d(q, t)$, whereas in Schrödinger dynamics, it is realized at the level of $\bar{\psi}(q, t)$.

At the same time, individual events must be a member of the entire distribution. However, to fully account for the relationship between "parts" and "the whole," a quantum stochastic dynamics description of individual particle motion is ultimately required. However, the motion of individual quantum particles cannot be treated in the same way as that of individual Brownian particles. This issue will be discussed in the next chapter and in Chap. 9.

7.2.3 Quantum Entanglement as a Quantum Superposition

In quantum theory, the principle of summation is usually referred to as the principle of superposition. This terminology reflects the historical interpretation of Schrödinger dynamics as a wave dynamics, leading to the notion of "wave superposition." Although the present text adopts the standpoint that the Schrödinger function is not a wavefunction but a distribution amplitude function, we shall follow convention and use the term "superposition principle."

The fact that the Schrödinger function itself represents a superposition of infinitely many events is made explicit by the path integral formulation. In this context, it is known that there are two types of superposed events: In the path sum of the Green function G, there are events that are expected to effectively vanish, as they do not play a physical role in the course of time evolution. On the other hand, for the physical and/or mathematical reasons, such as symmetry requirements in the Schrödinger equation, there exist quantum phenomena that can only be realized through specific combinations of two or more terms to form a proper irreducible representation. In such cases, the Schrödinger function is constructed as a superposition of independent events with fixed relative weights, and this structure must be preserved—it cannot disappear. This phenomenon, which has no counterpart in classical mechanics, is widely known as quantum entanglement. It was briefly discussed in Chap. 3, and more detail arguments will come in Sect. 9.5 of Chap. 9.

The paradoxes arising from the superposition principle in quantum theory are well known in the history of quantum mechanics [1]. Schrödinger himself was puzzled by his raised problem, now well known as "Schrödinger's cat," while Einstein was concerned with the issue of "spooky long-distance interactions faster than light." We briefly touch upon a comment on Schrödinger's cat first in this chapter and will discuss the entanglement in Chap. 9.

7.3 Density Matrix and Schrödinger's Cat

As noted above, the construction of a distribution function necessarily involves summation over events at some stage. In the case of the distribution amplitude function (the Schrödinger function), this is expressed as the principle of quantum mechanical superposition. However, it should be noted that this simple superposition principle strictly holds only for systems that are completely isolated from the external environment, i.e., perfectly closed systems. In this regard, insightful discussions have been given by Yutaka Toyozawa [2–4]. Before introducing these ideas, we briefly review the concept of the reduced density matrix.

7.3.1 Density Matrix

Resume that $\rho(q,t)$ was represented in the following form:

$$\rho(q,t) = \left(\phi_r(q,t)\ \phi_c(q,t)\right)\begin{pmatrix}\phi_r(q,t)\\ \phi_c(q,t)\end{pmatrix}. \tag{7.3.1}$$

One can make another matrix product such that

$$\begin{aligned}\hat{\rho}(q',q,t) &= \bar{\psi}(q,t)\bar{\psi}(q',t)^T\\ &\equiv \begin{pmatrix}\phi_r(q',t)\\ \phi_c(q',t)\end{pmatrix}\left(\phi_r(q,t)\ \phi_c(q,t)\right).\end{aligned} \tag{7.3.2}$$

In this case, $\hat{\rho}$ can be regarded as an operator. In the Dirac picture, $\hat{\rho}$ corresponds to

$$\hat{\rho}(t) \equiv |\psi(t)\rangle\,\langle\psi(t)|\,. \tag{7.3.3}$$

In a little more general form, we can extend it

$$\hat{\rho}_{ab}(t) \equiv |\psi_b(t)\rangle\,\langle\psi_a(t)|\,, \tag{7.3.4}$$

where $\hat{\rho}_{ab}(t)$ is called the density matrix operator. The trace operation for the matrix is

$$Tr\hat{\rho}_{ab}(t) = \langle\psi_a(t)\,|\psi_b(t)\rangle\,. \tag{7.3.5}$$

Since $\hat{\rho}(t)$ is an operator, it satisfies the Heisenberg equation of motion. That is

$$i\hbar\frac{d}{dt}\hat{\rho}(t) = \left[\hat{H},\hat{\rho}(t)\right]. \tag{7.3.6}$$

Do not misinterpret that the distribution function $\rho(q,t)$ is enough to track quantum dynamics. $\rho(q,t) = \langle q \left|\hat{\rho}(t)\right| q\rangle$ is not totally the same as $\hat{\rho}(t)$.

Next, let us define the reduced density matrix as follows:

$$\begin{aligned} Tr_{q_2}\langle q_1', q_2 \left|\hat{\rho}(t)\right| q_1, q_2\rangle &= \int dq_2 \langle q_1', q_2 \left|\hat{\rho}(t)\right| q_1, q_2\rangle \\ &\equiv \hat{\rho}_1(q_1', q_1, t). \end{aligned} \tag{7.3.7}$$

In this expression, the information of the q_2 space is averaged out and is no longer visible from the outside. However, this does not mean that the information from the q_2 space has been erased; rather, it means that the dynamics of the q_1 space are influenced by the effect from the q_2 space in average. Here, the function of the q_1 subsystem, which has received the influence from the q_2 subsystem, is denoted as $\hat{\rho}_1(q_1', q_1)$. Further, let us further diagonalize $\hat{\rho}_1(q_1', q_1, t)$ in such a manner that

$$\int dq_1 \hat{\rho}_1(q_1', q_1, t)\lambda_i^{(1)}(q_1, t) = n_i \lambda_i^{(1)}(q_1', t). \tag{7.3.8}$$

It is sometimes convenient to use the new set of functions $\left\{\lambda_i^{(1)}(q_1', t)\right\}$ [5,6]. These are called the natural orbitals.

The expectation value of an operator $\hat{A}$ with respect to $\rho(q,t)$ is given by

$$\left\langle \hat{A} \right\rangle = Tr \hat{A}\hat{\rho}(t) = \left\langle \psi(t) \left| \hat{A} \right| \psi(t) \right\rangle, \tag{7.3.9}$$

and this can be rewritten as

$$\begin{aligned} &Tr_{q_2}\left\langle q_1', q_2 \left| \hat{A}\hat{\rho}(t) \right| q_1, q_2 \right\rangle \\ &= \int dq_2 \psi^*(q_1', q_2, t)\hat{A}(q_1, q_2)\psi(q_1, q_2, t) \\ &\equiv \tilde{A}(q_1', q_1, t). \end{aligned} \tag{7.3.10}$$

The natural orbitals related to the expectation value provide the eigenvalues and eigenfunctions at time t in the q_1 subspace, which are influenced by the q_2 subspace. Such natural orbitals are extremely useful for understanding the phenomena in many-body problems [7–9].

7.3.2 The Principle of Superposition and Schrödinger's Cat

The paradox of "Schrödinger's cat" refers to the following thought experiment. A single cat, initially alive, is placed in a closed box. Inside the box, there is a device that releases a quantity of poisonous gas sufficient to kill the cat. In addition, a radioactive substance is placed in the box. Particles emitted by its spontaneous decay are converted into an electrical signal, which is then amplified so as to automatically trigger the gas-release mechanism. The decay of the radioactive substance is assumed to occur probabilistically over an appropriate time interval. If the quantum decay event occurs, the cat dies; if not, it remains alive. However, from outside the box, the state of the cat cannot be determined. Only after a sufficient time has elapsed—during which decay may or may not have occurred—and the box is opened, can the state of the cat be confirmed.

If so, prior to observation, the cat must be regarded as being in a superposition of being alive and dead. If quantum theory is correct, the wavefunction of the cat $\psi_{\text{cat}}(t)$ should be written as

$$|\psi_{\text{cat}}(t)\rangle = c_a(t)\,|\text{cat,alive}\rangle + c_d(t)\,|\text{cat,dead}\rangle\,, \tag{7.3.11}$$

where $|\text{cat,alive}\rangle$ and $|\text{cat,dead}\rangle$ denote the states in which the cat is alive and dead, respectively. The corresponding density matrix is

$$\begin{aligned}\hat{\rho}_{\text{cat}}(t) \equiv{}& |c_a(t)|^2\,|\text{cat,alive}\rangle\,\langle\text{cat,alive}| + |c_d(t)|^2\,|\text{cat,dead}\rangle\,\langle\text{cat,dead})| \\ &+ c_a(t)c_d^*(t)\,|\text{cat,alive}\rangle\,\langle\text{cat,dead}| + c_d(t)c_a^*(t)\,|\text{cat,dead}\rangle\,\langle\text{cat,alive}|\end{aligned} \tag{7.3.12}$$

In this expression, the ambiguity between life and death appears in the interference terms.

This paradox proposed by Schrödinger seems to highlight ambiguities in the interpretation of quantum theory, particularly those concerning the boundary between microscopic dynamics and macroscopic phenomena, as well as issues of human observability. Allegedly [10], he was inspired by discussions following the EPR paper [11] by Einstein and collaborators. Schrödinger himself is said to have been skeptical of the Copenhagen interpretation from the outset and later more sympathetic to Einstein's ensemble interpretation [10].

Before analyzing this paradox in detail, one should note that the setup of this problem itself is highly artificial. Biological phenomena are inherently high-level nonlinear processes and are therefore poorly suited as examples of the quantum superposition principle, which is fundamentally linear. Moreover, the timescales associated with radioactive decay and the determination of life or death are entirely different, and the distinction between "alive" and "dead" is not defined medicinally in a clear-cut manner and depends on the criteria adopted. From a physical standpoint, it would be preferable to replace "the life or death of the cat" with a more well-defined observable. Yet we proceed with this idealized scenario for the sake of discussion.

According to Toyozawa [4], the fundamental mistake lies in writing the wavefunction $\psi_{\text{cat}}(t)$ in the form of Eq. (7.3.11), as though the cat was completely isolated from its environment. The fate of the cat acquires meaning only within its environmental context. Therefore, at the very least, the state should be expressed as an entangled state with the radioactive decay process:

$$\left|\psi_{\rm sys}(t)\right\rangle = C_a(t)\,|{\rm cat,alive}\rangle\,|{\rm nondecayed}\rangle + C_d(t)\,|{\rm cat,dead}\rangle\,|{\rm decayed}\rangle\,. \tag{7.3.13}$$

Here, $|{\rm decayed}\rangle$ denotes the state in which the radioactive substance has decayed and emitted a particle, while $|{\rm nondecayed}\rangle$ denotes the nondecayed state. In principle, such system-environment linkages extend indefinitely to the surrounding outer systems, but for the present level of description, Eq. (7.3.13) suffices.

If one wishes to focus only on the classical status of the cat, as in Eq. (7.3.12), it is necessary to trace out the environmental degrees of freedom and construct a reduced density matrix. Assuming that the decayed and nondecayed states are orthogonal

$$\langle{\rm nondecayed}\,|{\rm decayed}\,\rangle = 0, \tag{7.3.14}$$

the reduced density matrix for the cat within the system is given by

$$\begin{aligned}&\hat{\rho}_{\text{cat-in-sys}}(t)\\&= Tr_{\rm rad}\left|\psi_{\rm sys}(t)\right\rangle\left\langle\psi_{\rm sys}(t)\right|\\&= |C_a(t)|^2\,|{\rm cat,alive}\rangle\,\langle{\rm cat,alive}| + |C_d(t)|^2\,|{\rm cat,dead}\rangle\,\langle{\rm cat,dead})|\,.\end{aligned} \tag{7.3.15}$$

$Tr_{\rm rad}$ indicates the trace over the subspace associated with the radioactive degrees of freedom. In this reduced density matrix, the state of the cat is represented as a statistical mixture with appropriate weights, and the apparent superposition between "alive" and "dead," as seen in Eq. (7.3.12), no longer appears. This demonstrates that, when analyzing the Schrödinger function of a subsystem, one must carefully account for its separability from the surrounding environment. Toyozawa emphasized that the problem of measurement in quantum mechanics requires consideration of such chains of interaction between a system and its surrounding environments [2–4]

For the same reason, in the quantum mechanics of systems embedded in an environment—such as dissipative systems, where the system is effectively convoluted with a heat bath—the appropriate starting point is the density matrix and its reduction. In the present text, however, we restrict attention to the quantum state of the subsystem before observation and do not pursue these issues further.

Meanwhile, what troubled Einstein was the persistence of superposition states maintained by entanglement, regardless of spatial separation. When entanglement is preserved between distant subsystems, peculiar phenomena are theoretically predicted. Then the natural question is: can't entanglement be broken and thereby resolved? If so, by what mechanism? Since the intrinsic symmetries of the Schrödinger equation cannot be violated, such disentanglement does not occur as long as one follows only the solutions of the Schrödinger equation. These issues will be revisited after constructing the theory of quantum stochastic paths in the next chapter and will form one of the central topics in Chap. 9.

References

1. Freire Jr, O. *et al. The Oxford Handbook of the History of Quantum Interpretations* (Oxford University Press, 2022).
2. Toyozawa, Y. Quantum logic, statistical operator and the problem of measurement. *Journal of the Physical Society of Japan* **53**, 3335–3341 (1984).
3. Toyozawa, Y. The role of statistical operator in quantum mechanics. *Journal of the Physical Society of Japan* **55**, 2572–2580 (1986).
4. Toyozawa, Y. Theory of measurement: a note on conceptual foundation of quantum mechanics. *Progress of Theoretical Physics* **87**, 293–305 (1992).
5. Löwdin, P.-O. Quantum theory of many-particle systems. I. Physical interpretations by means of density matrices, natural spin-orbitals, and convergence problems in the method of configurational interaction. *Phys. Rev.* **97**, 1474 (1955).
6. Davidson, E. R. Properties and uses of natural orbitals. *Rev. Mod. Phys.* **44**, 451 (1972).
7. Takatsuka, K. & Arasaki, Y. Energy natural orbitals. *J. Chem. Phys.* **154**, 094103 (2021).
8. Takatsuka, K. & Arasaki, Y. An orbital picture extracted from correlated electronic wavefunctions and application to forbidden reactions: 70 years of the frontier orbital theory. *J. Chem. Phys.* **155**, 064104 (2021).
9. Arasaki, Y. & Takatsuka, K. An orbital picture extracted from correlated electronic wavefunctions and application to forbidden reactions: 70 years of the frontier orbital theory. *J. Chem. Phys.* **156**, 234102 (2022).
10. Home, D. & Whitaker, A. *Einstein's Struggles with Quantum theory: a Reappraisal* (Springer Science & Business Media, 2007).
11. Einstein, A., Podolsky, B. & Rosen, N. Can quantum-mechanical description of physical reality be considered complete? *Physical Review* **47**, 777 (1935).

8 Quantum Stochastic Paths and Stochastic Differential Equations

Abstract

Contents in this chapter:

1. An Ito-type stochastic differential equation for quantum stochastic paths is given.
2. The solution of this stochastic differential equation formally resembles that of classical Brownian motion. Yet, the velocity field coefficient is a functional of the Schrödinger function.
3. The trajectories generated from the velocity field coefficient alone coincide with Bohm trajectories in the absence of the Wiener process and furthermore correspond to individual streamlines of the flux (manifold) generated by the real Schrödinger vector.
4. Basic properties of the quantum stochastic paths are shown.

Keywords

Quantum stochastic differential equation · Quantum dynamics of individual paths · Bohm trajectory Velocity field as a functional of Schrödinger function

In the preceding chapters, it has been shown that the real-valued Schrödinger equation—under a potential function—has a form closely analogous to that of the diffusion equation. Likewise, the real-valued quantum path integral has been shown to be quite akin to the Feynman-Kac formula in its mathematical structure. Consequently, it is expected that a certain correspondence should exist between the individual Schrödinger dynamics and the dynamics of Brownian particles. We investigate this aspect in this chapter.

K. Takatsuka, *Structure of the Schrödinger Dynamics*,
https://doi.org/10.1007/978-981-92-1910-0_8

First, we recall in the theory of stochastic processes, as discussed in Chap. 4, the following three elements that are mutually consistent and form a unified framework:

(1) The diffusion equation for the probability distribution function
(2) The Feynman-Kac formula for the Green function
(3) The Ito stochastic differential equation describing individual trajectories

In the Ito equation, the drift term of the velocity field is expressed in terms of the solution of the diffusion equation (see Eq. (4.4.8)), and therefore, unlike the Langevin equation, it does not constitute an independent equation of motion by itself. Therefore, the Ito stochastic differential equation cannot be constructed without the information of the density distribution function; hence, the description of individual trajectories does not form a self-contained structure. Although the resulting velocity field is consistent with the Feynman-Kac formula, the latter merely represents the sum of probabilities of all paths and has no function to dynamically generate individual trajectories.

As noted above, quantum dynamics exhibits a structure analogous to that of classical diffusion processes, even though they lie in the different hierarchical level. Meanwhile, for diffusion equations and the Fokker-Planck equation, there exist theorems due to Kolmogorov and Ito stating that a corresponding stochastic process underlies the parabolic partial differential equations. It is therefore natural to expect that an Ito-type stochastic differential equation may also underlie the Schrödinger dynamics behind the Schrödinger equation.

A significant first step in this direction was taken by Masao Nagasawa [1–3]. However, Nagasawa's formulation of quantum dynamics does not directly refer to the Schrödinger equation itself, as will be discussed later. Since both the real-valued Schrödinger equation and the real-valued quantum path integral have been derived in the present book, we next proceed to investigate the derivation and properties of quantum stochastic processes as one of the remaining fundamental pillars supporting the structure of Schrödinger dynamics

8.1 Concept of Quantum Stochastic Dynamical Path

There has long been interest in whether a physically meaningful path governed by dynamical laws in quantum mechanics exists, which might be analogous to classical trajectories. (We do not refer to semiclassical mechanics as the fundamental stochastic quantum dynamics, even though classical trajectory plays a main role there.)

In this context, the trajectory theory of David Bohm (see Eqs. (3.2.21) and (3.2.24)) is of particular importance. However, Bohmian trajectories correspond to streamlines of a fluid expressed by Eq. (3.2.21)—namely, trajectories of massless points flowing along the current—and it has been argued that they do not necessarily represent the intrinsic dynamical motion of quantum particles [4].

On the other hand, starting from stochastic processes and reconstructing the Schrödinger equation, Edward Nelson carried out pioneering work in the history of quantum theory [5,6]. Such an approach, in which one begins from an alternative theoretical foundation and arrives at the formal structure of quantum mechanics, is generally referred to as quantization. In other words, it is an attempt to uncover and establish the fundamental principles underlying quantum mechanics. The attempt by Schrödinger to derive his equation from the Hamilton-Jacobi theory (Eq. (3.2.18)) may also be regarded as a kind of quantization.

In contrast to Nelson's approach, Nagasawa constructed a stochastic process for trajectories starting from the standpoint of solving parabolic partial differential equations. The author considers Nagasawa's work and its perspective to be epoch-making in the history of quantum mechanics. In what follows, the essential features of these two approaches are first reviewed [1–3], after which the developments specific to the present work are presented.

8.1.1 Nelson's Stochastic Newton Equation

In the introduction to his paper, Nelson described the physical picture underlying his theory as follows [5] "If we have, for example, a hydrogen atom in the ground state, the electron is in dynamical equilibrium between the random force causing the Brownian motion and the attractive Coulomb force of the nucleus. Its trajectory is very irregular. Most of the time the electron is near the nucleus, sometimes it goes farther away, but it always shows a general tendency to move toward the nucleus, and this is true no matter which direction we take for time. This behavior is quite analogous to that of a particle in a colloidal suspension, in dynamical equilibrium between osmotic forces and gravity. However, the electron in the hydrogen atom has other states of dynamical equilibrium, at the usual discrete energy levels of the atom." Upon closer examination, however, it is not immediately clear what constitutes the "random force" acting on the electron in a hydrogen atom. This point requires careful clarification. It will be revisited briefly in Sect. 9.3 of Chap. 9. For the present, we proceed to the essential structure of Nelson's theory [5,6].

His theoretical formalism is somewhat involved as follows. Regarding the position of the particle $\mathbf{q}(t)$, we define the mean value of the differential using the following expectation value $E_t[\cdot]$: For the forward derivative

$$DX_t = \lim_{\Delta t \to +0} E_t \left[\frac{\mathbf{q}(t+\Delta t) - \mathbf{q}(t)}{\Delta t} \right] \tag{8.1.1}$$

and for the backward derivative

$$D_* \mathbf{X}_t = \lim_{\Delta t \to +0} E_t \left[\frac{\mathbf{q}(t) - \mathbf{q}(t-\Delta t)}{\Delta t} \right]. \tag{8.1.2}$$

These are called the forward velocity and backward velocity, respectively. Hereafter, for consistency with the notation in the latter half of this book, we will denote the stochastic variable $\mathbf{q}(t)$ as $\mathbf{X}_t$. Since $\mathbf{X}_t$ is continuous (not necessarily smooth though), the velocity of an ordinary Brownian particle coincides with

$$D\mathbf{X}_t = D_*\mathbf{X}_t \equiv \mathbf{v}(t) \tag{8.1.3}$$

The velocity is thus defined as

$$d\mathbf{X}_t = \mathbf{v}(t)dt \tag{8.1.4}$$

On the other hand, since $\mathbf{v}(t)$ is not continuous, the two-sided difference

$$d\mathbf{v}(t) = -\gamma\mathbf{v}(t)dt - \frac{1}{m}\frac{\partial V}{\partial \mathbf{q}}dt + d\mathbf{B}(t) \tag{8.1.5}$$

and

$$d\mathbf{v}(t) = \gamma\mathbf{v}(t)dt - \frac{1}{m}\frac{\partial V}{\partial \mathbf{q}}dt + d\mathbf{B}_*(t) \tag{8.1.6}$$

can be considered, corresponding to the Langevin equation. Here, γ is the friction coefficient, and $\mathbf{B}(t)$ and $\mathbf{B}_*(t)$ are Brownian (Wiener) processes. We assume that the external force $-\partial V/\partial \mathbf{q}$ is acting. When applying the differentials in Eqs. (8.1.1) and (8.1.2) to the Langevin equation (8.1.5), the mean value of the Brownian process differential vanishes such that

$$E_t(\frac{d\mathbf{B}}{dt}) = E_t(\frac{d\mathbf{B}_*}{dt}) = 0, \tag{8.1.7}$$

and it follows after all that

$$D\mathbf{v}(t) = -\gamma\mathbf{v}(t) - \frac{1}{m}\frac{\partial V}{\partial \mathbf{q}} \tag{8.1.8}$$

and

$$D_*\mathbf{v}(t) = \gamma\mathbf{v}(t) - \frac{1}{m}\frac{\partial V}{\partial \mathbf{q}}. \tag{8.1.9}$$

The first highlight of the theory is now seen; by using Eqs. (8.1.8) and (8.1.9), the following expression

$$\frac{1}{2}(DD_* + D_*D)\mathbf{X}_t = -\frac{1}{m}\frac{\partial V}{\partial \mathbf{q}} \tag{8.1.10}$$

has been formulated. This equation is what he calls the stochastic Newtonian equation. This is the starting point of the long (and rather winding) path to the Schrödinger equation.

Note that, in Eqs. (8.1.1) and (8.1.2), it is not explicitly shown what the expectation $E_t[\cdot]$ (conditional expectation) is taken with respect to, i.e., which distribution function is being used. Also, D_*D is not the expectation value of the direct second-order derivative, but rather, it is the expectation value taken after first differentiating forward and then backward. The same applies to DD_*.

To proceed further, we introduce the following notation:

$$d\mathbf{X}_t = \mathbf{b}(\mathbf{X}_t, t)dt + d\mathbf{W}(t), \tag{8.1.11}$$

where $\mathbf{b}(\mathbf{X}_t, t)$ is the velocity drift term, and

$$D\mathbf{X}_t = \mathbf{b}(\mathbf{X}_t, t). \tag{8.1.12}$$

Similarly, for the backward derivative, we have

$$d\mathbf{X}_t = \mathbf{b}_*(\mathbf{X}_t, t)dt + d\mathbf{W}_*(t) \tag{8.1.13}$$

and

$$D_*\mathbf{X}_t = \mathbf{b}_*(\mathbf{X}_t, t). \tag{8.1.14}$$

Below, the velocity drift term and related quantities will play a central role. The diffusion coefficient of this stochastic field will, in this particular section, be denoted by λ, and then, the Wiener process is represented as follows:

$$E_t\left(d\mathbf{W}_i d\mathbf{W}_j\right) = 2\lambda\delta_{ij}dt. \tag{8.1.15}$$

The diffusion constant λ is to be set to $\hbar/2m$ in quantum dynamics [5].

Before proceeding, we remind of Ito's formula for a given function: $f(x(t), t)$

$$Df(x(t), t) = (\partial/\partial t + \mathbf{b}\cdot\nabla + \lambda\nabla^2)f(x(t), t) \tag{8.1.16}$$

and

$$D_*f(x(t), t) = (\partial/\partial t + \mathbf{b}_*\cdot\nabla - \lambda\nabla^2)f(x(t), t), \tag{8.1.17}$$

which will be applied implicitly in what follows.

First, by substituting Eqs. (8.1.11) and (8.1.13) into the stochastic Newtonian equation (8.1.10), it can be rewritten as

$$DD_*\mathbf{X}_t = \left(\frac{\partial \mathbf{b}_*}{\partial t} + \mathbf{b}\cdot\nabla\mathbf{b}_* + \frac{\hbar}{2m}\nabla^2\mathbf{b}_*\right)\mathbf{X}_t \tag{8.1.18}$$

and

$$D_* D\mathbf{X}_t = \left(\frac{\partial \mathbf{b}}{\partial t} + \mathbf{b}_* \cdot \nabla \mathbf{b} - \frac{\hbar}{2m} \nabla^2 \mathbf{b} \right) \mathbf{X}_t. \tag{8.1.19}$$

Next, from the velocity fields $\mathbf{b}$ and $\mathbf{b}_*$, we define the following quantities related to the two types of velocities:

$$\mathbf{v} = \frac{1}{2}(\mathbf{b} + \mathbf{b}_*) \tag{8.1.20}$$

and

$$\mathbf{u} = \frac{1}{2}(\mathbf{b} - \mathbf{b}_*). \tag{8.1.21}$$

Using Eqs. (8.1.11) and (8.1.13), as well as Ito's formula, we obtain the following coupled (multidimensional) nonlinear differential equations for $\mathbf{u}$ and $\mathbf{v}$ such that

$$\partial \mathbf{u}/\partial t = -(\hbar/2m)\vec{\nabla}(\vec{\nabla} \cdot \mathbf{v}) - \vec{\nabla}(\mathbf{v} \cdot \mathbf{u}) \tag{8.1.22}$$

and

$$\partial \mathbf{v}/\partial t = (1/m)F - (\mathbf{v} \cdot \vec{\nabla})\mathbf{v} + (\mathbf{u} \cdot \vec{\nabla})\mathbf{u} + (\hbar/2m)\nabla^2 \mathbf{u}. \tag{8.1.23}$$

Assuming that these can be solved, we further reexpress $\mathbf{v}$ and $\mathbf{u}$ using the parameters S, R_N, and ρ as follows:

$$\mathbf{v} \equiv \frac{\hbar}{m} \nabla S \tag{8.1.24}$$

and

$$\mathbf{u} \equiv \frac{\hbar}{2m} \nabla \rho = \frac{\hbar}{m} \nabla R_N. \tag{8.1.25}$$

Here, ρ is supposed to satisfy the equation of continuity:

$$\frac{\partial}{\partial t} \rho = -\nabla \cdot (\mathbf{v}\rho) \tag{8.1.26}$$

with the above $\mathbf{v}$. And $\mathbf{u}$ turns out to be

$$\mathbf{u} \equiv \frac{\hbar}{2m\rho} \nabla \rho. \tag{8.1.27}$$

Equation (8.1.26) takes the same form as the classical continuity equation. (See also Eqs. (5.1.28) and (5.1.29).) However, as will be precisely discussed later, **u** of Eq. (8.1.27) is the velocity corresponding to $p^{\mathrm{Imag}}(q,t)$ defined in Eq. (5.2.30), and as we have shown above in Chap. 6, $p^{\mathrm{Imag}}(q,t)$ itself is not a dynamical variable in quantum mechanics.

To conclude the long journey, he defines the following function ψ in terms of S and R_N such that

$$\psi \equiv \rho^{1/2} \exp(iS) = \exp(R_N + iS). \tag{8.1.28}$$

(Distinguish this functional form from the expression of Bohm's one $\psi = R\exp(iS_B/\hbar)$ defined in Eq. (3.2.20).) In this way, the ψ in Eq. (8.1.28) is a complex Schrödinger function, and the Schrödinger equation

$$i\hbar\frac{\partial}{\partial t}\psi(q,t) = \left(-\frac{\hbar^2}{2m}\nabla^2 + V\right)\psi(q,t) \tag{8.1.29}$$

has appeared replacing the coupled nonlinear equations (8.1.22) and (8.1.23).

In reality, of course, the Schrödinger equation, which is a linear equation, must be much easier to handle than Eqs. (8.1.22) and (8.1.23), which are coupled nonlinear equations. By using Eqs. (8.1.20), (8.1.21), (8.1.24), and (8.1.25), the velocity field coefficient **b** and $\mathbf{b}_*$, respectively, can be rewritten in terms of the components of ψ in Eq. (8.1.28) as

$$\mathbf{b} = \frac{\hbar}{m}(\nabla S + \nabla R_N) \tag{8.1.30}$$

and

$$\mathbf{b}_* = \frac{\hbar}{m}(\nabla S - \nabla R). \tag{8.1.31}$$

The velocity field coefficient is one of the main topics of this chapter.

Through such intricate manipulations of equations, it is indeed possible to arrive at the Schrödinger equation starting from his stochastic Newtonian equation. Nelson argued that "the Schrödinger equation can be derived from the classical mechanics of Eq. (8.1.10)" and further claimed that "the Born probability interpretation emerges naturally in this framework." In fact, the title of his paper, "Derivation of the Schrödinger Equation from Newtonian Mechanics," [5]) strongly suggests that Newtonian mechanics itself sets the foundation of the Schrödinger equation.

By combining stochastic process theory with the Newtonian equation and subsequently quantizing the system to obtain the Schrödinger equation, this work represents a significant milestone in the history of quantum mechanics. Nevertheless, it should be noted that such a derivation relies heavily on formal transformations, the physical meaning of which is not necessarily evident. For instance, if one attempts to reverse the procedure with the use of identical back transformations—namely, to recover Eq. (8.1.10) from Eq. (8.1.29)—the genuine classical Newtonian equation for particles is unlikely to be obtained by limiting $\hbar \to 0$.

On the other hand, as discussed in Chap. 5 of this book, the real-valued Schrödinger equation could be derived from first principles, and its physical meaning was made explicit. From this perspective, Nelson's quantization procedure appears somewhat indirect, and moreover, several aspects of the theory remain difficult to interpret physically. Indeed, if quantization alone is the objective, such a reconstruction is not necessary in view of our straightforward derivation. Yet some readers may find that Nelson's original paper [5] provides additional insight into the Schrödinger equation and its associated interpretation. However, it seems that few genuinely new physical phenomena, approximation schemes, or interpretations have emerged from the coupled nonlinear equations (8.1.22) and (8.1.23).

Kunio Yasue, by developing his stochastic variational method, reformulated the Nelson framework by discarding the stochastic Newtonian equation (8.1.10) and instead starting from the following action functional:

$$\frac{d}{dt}\int\left[\frac{1}{2}(\frac{m}{2}\mathbf{b}^2+\frac{m}{2}\mathbf{b}_*^2)+V(q)\right]\rho(q,t)=0. \qquad (8.1.32)$$

Interestingly, he treated the problem within the framework of optimal control theory, from which he has shown the Schrödinger equation can be derived and accordingly interpreted [7,8]. Equation (8.1.32) can be understood, in a direct sense, as expressing the conservation of classical energy within a probabilistic setting. The treatment of stochastic differential equations proceeds in parallel with that of Nelson. Notably, his finding a variational principle of control theory underlying quantum mechanics constitutes an important conceptual advance.

The quantization method of Nelson and related approaches are sometimes referred to as stochastic quantization. However, these should be distinguished from the stochastic quantization proposed by Parisi and Wu [9, 10]. In the Parisi-Wu framework, a fictitious time variable is introduced in physical space-time in order to describe statistical ensembles. A stochastic process is then defined with respect to this auxiliary time, and the equilibrium limit in this extended space reproduces ordinary quantum theory.

In addition, there exist formulations such as the stochastic Schrödinger equation, in which stochastic processes are introduced to the Schrödinger equation either as external fields or as effective descriptions of interactions with an environment (e.g., a heat bath) in dissipative systems. By contrast, the concept of quantum stochastic trajectories considered in this chapter is based on the assumption that an intrinsic stochastic process should underlie the Schrödinger dynamics itself. It is therefore fundamentally different from these approaches. The reader should keep this distinction in mind. Although the stochastic Schrödinger equation is an interesting subject, we shall not pursue the relevant details further in this book.

8.1.2 Theory of Nagasawa's Quantum Path Stochastic Differential Equation

As outlined above, Nelson's program begins by postulating the stochastic differential equations (8.1.11) and (8.1.13). These are then substituted into the stochastic Newtonian equation (8.1.10), and through appropriate transformations of variables, the Schrödinger equation is ultimately obtained.

In contrast, the theory developed by Nagasawa is based on the mathematically rigorous result that partial differential equations of parabolic-type like diffusion equations and Fokker-Planck equations are always associated with relevant stochastic differential equations. (Refer also to Chap. 4, Sect. 4.4.3 entitled "Relationship between the Parabolic Partial Differential Equation and the Stochastic Process.") Building on this fact, Nagasawa formulates path-wise stochastic differential equations together with a system of coupled real-valued partial differential equations that serve as an alternative formulation of the Schrödinger equation. In this framework, stochastic processes are not assumed a priori; rather, they emerge as a consequence of the theory.

A complete reproduction of his great work lies beyond the scope of the present book, and only a brief outline is provided here. Readers interested in further details are encouraged to consult the original works [1, 3]. Nagasawa has also published expositions in Japanese—for example, New Quantum Theory Based on the Markov Process Theory (Masao Nagasawa, Soeisha/Sanseido, 2012)—which may serve as a more accessible reference.

Let us start from the following fact. Suppose there exists a Markov process that drives a function $u(x, t)$ to evolve in space-time with a transition probability density function $q(t, x, 0, y)$ for any initial function $f(y)$ as

$$u(x,t) = \int q(t, x, 0, y) f(y) dy. \tag{8.1.33}$$

If this $u(x, t)$ satisfies a parabolic-type partial differential equation (diffusion equation or, more generally, the Fokker-Planck equation) as

$$\frac{\partial u}{\partial t} + \frac{1}{2}\sigma^2 \nabla^2 u + \mathbf{b}(q, t) \cdot \nabla u = 0, \tag{8.1.34}$$

then this Markov process is equivalent to the stochastic process represented as

$$X_t = X_0 + \sigma W_t + \int_0^t \mathbf{b}(X_s, s) ds. \tag{8.1.35}$$

In this context, the constant σ and the function $\mathbf{b}$ are common to those in Eqs. (8.1.34) and (8.1.35), respectively. Also, W_t is a Wiener (Brownian) process. The function $\mathbf{b}(q, t)$ can include the vector potential of an electromagnetic field, but in the following, we set this particular part to zero. (To avoid confusion, note that the $\mathbf{b}(q,t)$ here is not the $\mathbf{b}$ in Nelson's theory, Eq. (8.1.12).) Using this, we want to derive the partial differential equation behind the stochastic paths underlying quantum dynamics. Unfortunately, however, the Schrödinger equation does not take the form of Eq. (8.1.34) and is a partial differential equation with complex coefficients. Therefore, we need something else to attain a stochastic integral or differential equation like Eq. (8.1.35). Nagasawa came to the following theory after his careful and deep study.

Suppose that a Schrödinger function is represented as

$$\psi(q,t) = \exp\left(R_N(q,t) + iS(q,t)\right) \tag{8.1.36}$$

under a potential function $V(q,t)$. Here, both $R_N(q,t)$ and $S(q,t)$ are assumed to be real-valued functions. From Eq. (8.1.36), we may define two real-valued functions

$$\psi_N(q,t) = \exp(R_N(q,t) + S(q,t)) \tag{8.1.37}$$

and

$$\tilde{\psi}_N(q,t) = \exp(R_N(q,t) - S(q,t)) \tag{8.1.38}$$

After somewhat lengthy manipulation, Nagasawa has shown that each of them satisfies

$$\hbar\frac{\partial\psi_N}{\partial t} + \frac{1}{2}\sigma^2\left(\nabla + \mathbf{b}(q,t)\right)^2\psi_N + c(q,t)\psi_N = 0 \tag{8.1.39}$$

and

$$-\hbar\frac{\partial\tilde{\psi}_N}{\partial t} + \frac{1}{2}\sigma^2\left(\nabla - \mathbf{b}(q,t)\right)^2\tilde{\psi}_N + c(q,t)\tilde{\psi}_N = 0, \tag{8.1.40}$$

where $c(q,t)$ is defined by

$$c(q,t) = -V(q,t) - 2\frac{\partial S}{\partial t} - \sigma^2\left(\nabla S\right)^2 - 2\sigma^2\mathbf{b}\cdot\nabla S - \sigma^2\mathbf{b}^2 \tag{8.1.41}$$

and σ^2 is given by

$$\sigma^2 = \frac{\hbar}{m}. \tag{8.1.42}$$

The proof for this fact is somewhat involved, and those who wish to study this in depth are referred to the papers [1,3] and the previous book *A New Quantum Theory Based on Markov Processes*.

For the parabolic partial differential equation (8.1.39), we accordingly obtain the corresponding stochastic differential equation as

$$dX_t = \frac{\hbar}{m}\nabla(R_N + S)dt + \sqrt{\frac{\hbar}{m}}dW_t. \tag{8.1.43}$$

Here, it turns out that the velocity field is explicitly given by

$$\frac{\hbar}{m}\nabla(R_N + S), \tag{8.1.44}$$

the components of which can be attained from the information of the Schrödinger function. Note however that ψ_N is not the Schrödinger equation itself. Nevertheless, the velocity field in Eq. (8.1.44) is exactly the same as that of Nelson of Eq. (8.1.30). In other words, as long as the same Schrödinger function is used, these two stochastic differential equations are identical.

As can be seen, Eq. (8.1.40) for $\tilde{\psi}_N$ is the time-reversed version of the evolution equation (8.1.39) for ψ_N. Therefore, in order to solve for $\left\{\psi_N, \tilde{\psi}_N\right\}$ simultaneously, it is necessary to use not only the initial condition but also the final one. The coupled equations (8.1.39) and (8.1.40) are not, strictly speaking, Schrödinger equations themselves, but once the set $\left\{\psi_N, \tilde{\psi}_N\right\}$ is obtained, the Schrödinger function can be reproduced inversely via Eq. (8.1.36). Furthermore, Nagasawa claims that, since the coupled equations (8.1.39) and (8.1.40) possess the properties that the Schrödinger equation does not, his formalism goes beyond quantum mechanics. The details are profound, but in this book, we will stop here and refer the interested reader to Nagasawa's works [1, 3].

8.2 Quantum Trajectories from the Schrödinger Equation as the Coupled Diffusion Equations

In Chap. 4, we study the basics of the stochastic process dynamics, namely, the diffusion equation, the Feynman-Kac path integral formula, and the Ito stochastic differential equation. We pointed out that, in the process from the Schrödinger equation to the real-valued path integral, the Schrödinger equation in Eq. (6.2.2) is akin to the diffusion equation in Eq. (4.3.7). Below, based on this observation, we then further attempt to determine a stochastic differential equation for the quantum stochastic path [11].

Here, we are motivated to reconsider the stochastic differential equation for the quantum path by the following issues. One is that, in both Nelson's theory and Nagasawa's theory, the velocity field $\frac{\hbar}{m}\nabla(R_N + S)$ contains the term ∇R_N. Therefore, the velocity field is expressed, using both the real and imaginary parts R_N and S of the exponent in the complex Schrödinger function $\psi = \exp(R_N + iS)$ of Eq. (8.1.36) as in Eqs. (8.1.30) and (8.1.44). This raises the following fundamental questions. To study this aspect, we tentatively focus only on the velocity field in the stochastic differential equation and deliberately ignore the Wiener process for the sake of discussion.

(1) The trajectories to be attained by integrating the stochastic differential equations (with no Wiener process) with the Nelson-Nagasawa velocity field are different from those of the Bohm trajectory. This is because the Bohm trajectory is determined only by $\frac{1}{m}\nabla S_B$ [writing $\psi(q,t) = R(q,t)\exp\left(\frac{i}{\hbar}S_B(q,t)\right)$], but not $\frac{\hbar}{m}\nabla(R_N + S)$. Although the Bohm trajectory is certainly the motion of the flow line induced by the quantum action field S_B, we may regard it as being directly related to the motion of particles, since it directly represents the movement and deformation of the Schrödinger distribution amplitude function.

(2) ∇R_N, by using the density function $\rho = \exp(2R_N)$ (note again that $\psi = \exp(R_N + iS)$), is expressed as

$$\nabla R_N = \frac{1}{2\rho}\nabla\rho. \tag{8.2.1}$$

However, the velocity proportional to $\nabla\rho$, in contrast to ∇S, is a non-quantum mechanical quantity, as seen in Eq. (5.2.30). $\nabla\rho$ appears as the flux in classical diffusion, as seen in Eq. (4.2.2). This is because the velocity component $\nabla\rho$ is induced only by the semigroup time evolution of the Schrödinger vector. It simply ignores the effect of vector rotation of the Schrödinger vector to be induced by the symplectic unit matrix $\mathbf{J}$ in time evolution (see Eq. (5.2.37)).

(3) ∇R_N does not have the correct time-reversal symmetry. While $S(-t) = -S(t)$, $R_N(-t) \neq -R_N(t)$. Therefore, $\frac{\hbar}{m}\nabla(R_N + S)$ does not have the symmetry of time reversal either.

(4) In Chap. 9 of this book, we will derive classical trajectory dynamics (the Hamilton canonical equations of motion and the Newtonian equation) from quantum trajectory dynamics. It will be shown then that the existence of ∇R_N acts as an extra term to hamper the reduction to classical dynamics. The aforementioned lack of time-reversal symmetry also prevents quantum stochastic dynamics from being reduced to classical counterpart.

(5) Furthermore, in Chap. 9 too, we will discuss the concept of the entrance and exit of quantum entanglement channels. Again, it will be shown that the existence of ∇R_N blocks the discussion of spontaneous disentanglement.

With these in mind, let us consider the stochastic differential equation for quantum trajectories directly connected to the Schrödinger equation, referring to the discussion of stochastic processes in Chap. 4 [11].

Revisiting the Ito Stochastic Differential Equation in Diffusion Dynamics

Before considering the quantum stochastic differential equation, let us briefly review the diffusion equation from the viewpoint of Schrödinger dynamics and the Ito stochastic differential equation in the context of diffusion. The diffusion equation in Eq. (4.3.7) is rewritten as

$$\frac{\partial \psi_D(q,t)}{\partial t} = \left(D\nabla^2 - \lambda V(q,t)\right)\psi_D(q,t), \tag{8.2.2}$$

and the corresponding stochastic differential equation is

$$dX_t = \alpha(X_t,t)dt + dW(t,\omega), \tag{8.2.3}$$

where the velocity field coefficient $\alpha(X_t,t)$ is given, using the solution $\psi_D(q,t)$ of the above diffusion equation, by

$$\alpha(X_t,t) = 2D\frac{\nabla\psi_D}{\psi_D}. \tag{8.2.4}$$

Thus

$$dX_t = 2D\frac{\nabla\psi_D}{\psi_D}dt + dW(t,\omega) \tag{8.2.5}$$

is the stochastic differential equation for diffusion.

On the other hand, the real Schrödinger equation can be written using the Schrödinger vector as

$$\frac{\partial}{\partial t}\begin{pmatrix}\phi_r(q,t)\\ \phi_c(q,t)\end{pmatrix} = \left(\frac{\hbar}{2m}\nabla^2 - \frac{V(q)}{\hbar}\right)\mathbf{J}\begin{pmatrix}\phi_r(q,t)\\ \phi_c(q,t)\end{pmatrix} \tag{8.2.6}$$

or, alternatively, as

$$\frac{\partial}{\partial t}\bar{\psi}(q,t) = \left(\frac{\hbar}{2m}\nabla^2 - \frac{1}{\hbar}V(q)\right)\mathbf{J}\bar{\psi}(q,t). \tag{8.2.7}$$

It can be confirmed that Eq. (8.2.7) is highly analogous to Eq. (8.2.2). Moreover, we notice that the quantum velocity field $\vec{p}^{\text{local}}(q,t)$ given in Eq. (5.2.15) has the form in terms of the complex Schrödinger function as

$$\vec{p}^{\text{local}}(q,t) = \hbar\,\text{Im}\,\frac{\vec{\nabla}\psi}{\psi}, \tag{8.2.8}$$

which is similar to Eq. (8.2.4). Therefore, a naive comparison between Eqs. (8.2.4) and (8.2.8) tempts us to deduce a relation that

$$\alpha(X_t,t) = 2D\frac{\nabla\psi_D}{\psi_D} \rightarrow \frac{\hbar}{m}\,\text{Im}\,\frac{\nabla\psi}{\psi} = \frac{\hbar}{m}\nabla S_B, \tag{8.2.9}$$

again assuming $\psi(q,t) = R(q,t)\exp\left(\frac{i}{\hbar}S_B(q,t)\right)$ in Bohm's representation. Besides, we have already seen the relation $D \leftrightarrow \hbar/2m$ in Eqs. (3.2.33) and (5.2.33). The final ∇S_B in Eq. (8.2.9) is the same as the $\hbar\nabla S$ in Nelson-Nagasawa. This intuitive assignment of Eq. (8.2.9) is in fact found to be correct. However, actually, we need a more careful discussion.

Quantum Trajectory

Even though Eq. (8.2.7) resembles the diffusion equation, it is not exactly, because of the presence of $\mathbf{J}$ that couples two independent equations. Consequently, the information contained within is more involved. In diffusion dynamics, the time evolution is determined by the semigroup operator, whereas for the actual Schrödinger equation, the evolution is unitary in the space of Schrödinger vectors, so its velocity field may not be fully described with a single real-valued coefficient $\alpha(X_t,t)$. Therefore, let us first diagonalize the $\mathbf{J}$ of Eq. (8.2.6) to exclude the formal coupling. This procedure simply brings us back to a pair of the complex Schrödinger equations

$$i\frac{\partial}{\partial t}(\phi_r + i\phi_c) = \left(\frac{\hbar}{2m}\nabla^2 - \frac{V}{\hbar}\right)(\phi_r + i\phi_c) \tag{8.2.10}$$

and

$$-i\frac{\partial}{\partial t}(\phi_r - i\phi_c) = \left(\frac{\hbar}{2m}\nabla^2 - \frac{V}{\hbar}\right)(\phi_r - i\phi_c). \tag{8.2.11}$$

Next, first, rewriting (rotating) the time t in Eq. (8.2.10) as

$$s^+ = -it, \tag{8.2.12}$$

we formally return to the diffusion equation of the form of Eq. (8.2.2), that is

$$\frac{\partial}{\partial s^+}\psi^+(q,s^+) = \left(\frac{\hbar}{2m}\nabla^2 - \frac{V}{\hbar}\right)\psi^+(q,s^+). \tag{8.2.13}$$

Here we have defined $\psi^+ = \phi_r + i\phi_c$. Similarly, for the Schrödinger equation (8.2.11), another formal transformation

$$s^- = it \tag{8.2.14}$$

along with $\psi^- = \phi_r - i\phi_c$ give

$$\frac{\partial}{\partial s^-}\psi^-(q,s^-) = \left(\frac{\hbar}{2m}\nabla^2 - \frac{V}{\hbar}\right)\psi^-(q,s^-). \tag{8.2.15}$$

Obviously, the functions $\psi^+(q,s^+)$ and $\psi^-(q,s^-)$ are mutually complex conjugate functions.

Formally regarding Eq. (8.2.13) as a diffusion equation in the "time axis" s^+, we see its velocity field coefficient be

$$\alpha^+(X_t, s^+)ds^+ = \frac{\hbar}{m}\frac{\nabla\psi^+}{\psi^+}ds^+, \tag{8.2.16}$$

according to Eq. (8.2.4). Likewise, from Eq. (8.2.15) in the time axis s^-, we have

$$\alpha^-(X_t, s^-)ds^- = \frac{\hbar}{m}\frac{\nabla\psi^-}{\psi^-}ds^-. \tag{8.2.17}$$

Obviously, $\alpha^+(X_t, s^+)ds^+$ and $\alpha^-(X_t, s^-)ds^-$ are mutually complex conjugates. We then rewrite them back in the time axis t in the conjugate form as

$$\alpha^+(X_t, s^+)ds^+ = \left[\alpha^{\text{Real}}(X_t, t) + i\alpha^{\text{Imag}}(X_t, t)\right]dt \tag{8.2.18}$$

and

$$\alpha^-(X_t, s^-)ds^- = \left[\alpha^{\text{Real}}(X_t, t) - i\alpha^{\text{Imag}}(X_t, t)\right]dt, \tag{8.2.19}$$

where both $\alpha^{\text{Real}}(X_t, t)$ and $\alpha^{\text{Imag}}(X_t, t)$ are real functions. $\alpha^{\text{Real}}(X_t, t)$ can be calculated as follows:

$$\begin{aligned}\alpha^{\text{Real}}(X_t, t)dt &= \frac{1}{2}\frac{\hbar}{m}\left(\frac{\nabla\psi^+}{\psi^+}ds^+ + \frac{\nabla\psi^-}{\psi^-}ds^-\right)\\ &= \frac{\hbar}{2m}\left(\frac{\nabla\psi^+}{\psi^+}(-idt) + \frac{\nabla\psi^-}{\psi^-}(idt)\right)\end{aligned} \tag{8.2.20}$$

and can be represented in terms of the real valued Schrödinger vector as

$$\begin{aligned}&\alpha^{\text{Real}}(X_t, t)dt\\ &= \frac{\hbar}{\rho m}dt\,(\phi_r\nabla\phi_c - \phi_c\nabla\phi_r)\\ &= -\frac{\hbar}{\rho m}\,(\phi_r\ \phi_c)\,\mathbf{J}\nabla\begin{pmatrix}\phi_r\\ \phi_c\end{pmatrix}dt.\end{aligned} \tag{8.2.21}$$

Here again, $\rho = \phi_r^2 + \phi_c^2$. Since the momentum operator is $\hat{p} = -\mathbf{J}\hbar\nabla$, α^{Real} is nothing other than the local momentum $\vec{p}^{\text{local}}(q, t)$ in Eq. (5.2.15) divided by the mass m. Therefore, from Eq. (5.2.15), we may write as

$$\alpha^{\text{Real}}(X_t, t) = \frac{\hbar}{m}\,\text{Im}\,\frac{\vec{\nabla}\psi}{\psi} \tag{8.2.22}$$

with the complex Schrödinger function, and this is exactly what we anticipated in Eq. (8.2.9). In order to summarize the relationship between $\alpha^{\text{Real}}(X_t, t)$ and $\vec{p}^{\text{local}}(q, t)$, we display $\vec{p}^{\text{local}}(q, t)$ here again

$$\begin{aligned}\vec{p}^{\text{local}}(q, t) &= \frac{\hbar}{\rho}\left(\phi_r \vec{\nabla}\phi_c - \phi_c \vec{\nabla}\phi_r\right)\\ &= \text{Re}\,\frac{\vec{p}\psi}{\psi} = \hbar\,\text{Im}\,\frac{\vec{\nabla}\psi}{\psi}.\end{aligned} \tag{8.2.23}$$

On the other hand, $\alpha^{\text{Imag}}(X_t, t)$ is calculated as follows:

$$\begin{aligned}\alpha^{\text{Imag}}(X_t, t)dt &= \frac{1}{2i}\frac{\hbar}{m}\left(\frac{\nabla\psi^+}{\psi^+}ds^+ - \frac{\nabla\psi^-}{\psi^-}ds^-\right)\\ &= -\frac{\hbar}{m\rho}dt\left(\phi_r\left(\nabla\phi_r\right) + \phi_c\left(\nabla\phi_c\right)\right)\\ &= -\frac{\hbar}{2m\rho}\left(\nabla\rho\right)dt.\end{aligned} \tag{8.2.24}$$

As for the physical meaning of $\alpha^{\text{Imag}}(X_t, t)$, we will consider it later.

Since the velocity field coefficients have appeared in a pair of quantity mutually complex conjugate, the relevant trajectory dX_t is also expressed as a complex quantity as

$$dX_t = dX_t^{\text{Real}} + i\,dX_t^{\text{Imag}}. \tag{8.2.25}$$

However, since we are interested in the real-valued trajectory dX_t^{Real}, which bears $\alpha^{\text{Real}}(X_t, t)$, we will rewrite dX_t^{Real} as the physically desired dX_t from now on, that is

$$\begin{aligned}dX_t &\equiv dX_t^{\text{Real}}\\ &= \alpha^{\text{Real}}(X_t, t)dt + \sqrt{\frac{\hbar}{m}}dW_0(t, \omega)\\ &= -\frac{\hbar}{\rho m}\left(\phi_r\ \phi_c\right)\mathbf{J}\nabla\begin{pmatrix}\phi_r\\ \phi_c\end{pmatrix}dt + \sqrt{\frac{\hbar}{m}}dW_0(t, \omega)\\ &= \frac{1}{\rho m}\left(\phi_r\ \phi_c\right)\hat{p}\begin{pmatrix}\phi_r\\ \phi_c\end{pmatrix}dt + \sqrt{\frac{\hbar}{m}}dW_0(t, \omega).\end{aligned} \tag{8.2.26}$$

Here, $W_0(t, \omega)$ is a standard Wiener process:

$$dW_0(t, \omega)\,dW_0(t, \omega) = dt \tag{8.2.27}$$

which is normalized as above. The corresponding diffusion coefficient D is, as previously given

$$D = \frac{\hbar}{2m}. \tag{8.2.28}$$

Thus, we have obtained the stochastic differential equation for the quantum stochastic path moving in real space. It is natural to consider that the Wiener process included here is the physical root behind Born's probability interpretation. Also, it should be added that the obtained $\alpha^{\text{Real}}(X_t, t)$ is invariant under the rotation of the Schrödinger vector.

On the Imaginary Velocity Field

Here, let us consider the meaning of $\alpha^{\text{Imag}}(X_t, t)dt$ as the counterpart to $\alpha^{\text{Real}}(X_t, t)dt$. Returning back to Eq. (8.2.18)

$$\begin{aligned} \alpha^+(X_t, s^+)ds^+ &= \left[\alpha^{\text{Real}}(X_t, t) + i\alpha^{\text{Imag}}(X_t, t)\right] dt \\ &= \alpha^{\text{Real}}(X_t, t)dt + \alpha^{\text{Imag}}(X_t, t)(idt), \end{aligned} \tag{8.2.29}$$

we see that $\alpha^{\text{Imag}}(X_t, t)$ can be interpreted as the velocity field for the stochastic process that accompanies the Schrödinger equation when the time variable is changed from t to it. By the replacement $t \to it$, the Schrödinger equation gives rise to

$$\frac{\partial}{\partial t}\psi(q, t) = \left(\frac{\hbar}{2m}\nabla^2 - \frac{1}{\hbar}V\right)\psi(q, t), \tag{8.2.30}$$

which is exactly the same as the original diffusion equation (8.2.2). It is also the same as the real Schrödinger equation (8.2.6), if $\mathbf{J}$ is replaced by the unit matrix to remove the coupling of two component equations (8.2.6).

Here, let us consider again the form of the local space-time distribution of the momentum variable, $\vec{p}\psi/\psi$. That is

$$\begin{aligned} \frac{\vec{p}\psi}{\psi} &= \frac{\hbar}{i}\frac{\vec{\nabla}\psi}{\psi} \\ &= \frac{\hbar}{i}\frac{1}{\rho}(\phi_r\nabla\phi_r + \phi_c\nabla\phi_c) + \frac{\hbar}{\rho}(\phi_r\nabla\phi_c - \phi_c\nabla\phi_r) \\ &\equiv ip^{\text{Imag}}(q, t) + p^{\text{local}}(q, t) \end{aligned} \tag{8.2.31}$$

so the quantum mechanical momentum is

$$p^{\text{local}}(q, t) = \frac{\hbar}{\rho}(\phi_r\nabla\phi_c - \phi_c\nabla\phi_r), \tag{8.2.32}$$

which appeared in $\alpha^{\text{Real}}(X_t, t)$, and the imaginary part of Eq. (8.2.31) is

$$\vec{p}^{\text{Imag}}(q, t) = -\frac{\hbar}{2\rho}\left(\vec{\nabla}\rho\right), \tag{8.2.33}$$

which corresponds to $\alpha^{\text{Imag}}(X_t, t)$. Thus, it turns out that $\vec{p}^{\text{Imag}}(q, t)$ is equivalent to $p^{\text{local}}(q, t)$ if $\mathbf{J}$ is replaced by $\mathbf{I}$ in Eq. (5.2.15), that is

$$\begin{aligned}
\vec{p}^{\text{Imag}}(q, t) &= \frac{1}{\rho}\left(\phi_r(q, t)\ \phi_c(q, t)\right)(-\mathbf{I}\hbar\vec{\nabla})\begin{pmatrix}\phi_r(q, t)\\ \phi_c(q, t)\end{pmatrix} \\
&= -\frac{\hbar}{\rho}\left(\phi_r\vec{\nabla}\phi_r + \phi_c\vec{\nabla}\phi_c\right) \\
&= -\frac{\hbar}{2\rho}\left(\vec{\nabla}\rho\right).
\end{aligned} \tag{8.2.34}$$

This is also the reproduction of Eq. (5.2.30). Furthermore, it is readily confirmed that $\vec{p}^{\text{Imag}}(q, t)$ is the same as $\vec{\nabla}R_N$ appearing in the Nelson-Nagasawa theory, Eqs. (8.1.30) and (8.1.44), except for the factor $-\hbar$.

In summary

$$\alpha^{\text{Real}}(X_t, t) = \frac{1}{m}p^{\text{local}}(X_t, t) \tag{8.2.35}$$

and

$$\alpha^{\text{Imag}}(X_t, t) = \frac{1}{m}p^{\text{Imag}}(X_t, t). \tag{8.2.36}$$

As described above, the operators $\exp(c_p\mathbf{J}\vec{\nabla}_q)$ with Eq. (5.1.13) and $\exp(c_t\mathbf{J}\partial_t)$ with Eq. (5.1.26) are space-time evolution operators that preserve unitarity (assuming c_p and c_t are real numbers). However, if we replace $\mathbf{J}$ with the identity matrix, the operators $\exp(c_p\vec{\nabla}_q)$ and $\exp(c_t\partial_t)$ become, for each, an evolution operator of a semigroup that does not consider vector rotation with respect to the Schrödinger vector. Therefore, comparing $\vec{p}^{\text{local}}(q, t)$ with $\mathbf{J}$ and $\vec{p}^{\text{Imag}}(q, t)$ without $\mathbf{J}$, we understand the necessity of $\mathbf{J}$ for the unitarity, including vector rotation, in the shift operations of Eqs. (5.1.6) and (5.1.22). In other words, if, instead of descending to the level of the Schrödinger vector and imposing unitarity on the spatiotemporal shift of $\rho(q, t)$, we relax the condition to a semigroup-type spatiotemporal evolution; then, $\vec{\nabla}\rho(q, t)$ and $\partial_t\rho(q, t)$ appear explicitly in the dynamics. This reveals an essential aspect of quantum theory. Furthermore, from Eqs. (5.2.34) and (8.2.31), it becomes clear that quantum theory and classical diffusion theory stand in a complementary (dual) relationship.

Nevertheless, even in quantum theory, it can happen for imaginary time or complex momentum to become necessary. This arises, for example, in the treatment of dynamics of nonconservative (dissipative) systems [12] or in dynamics under boundary conditions that render the Hamiltonian non-Hermitian (particularly in scattering theory), i.e., in so-called non-Hermitian quantum mechanics. However, we shall not pursue these topics further in this book.

Nelson and Nagasawa Revisited

Finally, let us return to the theories of Nelson and Nagasawa and re-examine their velocity field coefficients. In Nelson's theory, returning to Eqs. (8.1.11) and (8.1.13) and taking their averages as was done when writing down the stochastic Newton equation (8.1.10), we obtain

$$\frac{1}{2}(D + D_*)\mathbf{X}_t = \frac{1}{2}(\mathbf{b} + \mathbf{b}_*) = \frac{\hbar}{m}\nabla S \tag{8.2.37}$$

Here, Eqs. (8.1.30) and (8.1.31) have been used. Thus, the velocity field coefficient reduces to $\frac{\hbar}{m}\nabla S$.

In Nagasawa's theory, we need to return to Eq. (8.1.43). The velocity field coefficient in this expression originates from $\psi_N(q,t)$ in Eq. (8.1.37) and its equation of motion (8.1.39). On the other hand, from the time-reversed expression (8.1.38) with $\tilde{\psi}_N(q,t)$ and its equation of motion (8.1.40), we should obtain

$$dX_{-t} = \frac{\hbar}{m}\nabla(R_N - S)d(-t) + \sqrt{\frac{\hbar}{m}}dW_{-t}. \tag{8.2.38}$$

Taking the average of dX_t and dX_{-t} again leads to the reduction of the velocity field coefficient to $\frac{\hbar}{m}\nabla S$ in this case as well.

It is considered that the treatment of time in the author's Eq. (8.2.29) shares a fundamental commonality with those of Nelson and Nagasawa.

8.3 The Three Pillars of Schrödinger Dynamics and Dual Structure

In Chap. 4, regarding diffusion phenomena, it was introduced that the general diffusion equation, the Feynman-Kac path integral formula, and Ito stochastic differential equations form three pillars that constitute the foundation of the theory in a mutually related manner. On the other hand, for Schrödinger quantum dynamics, although further investigation remains, we have found that there exists an analogous tripartite structure corresponding to diffusion theory. These are the real-valued Schrödinger equation (6.2.1), the real-valued path integral as its Green function (6.2.6), and the stochastic differential equation for quantum stochastic paths (8.2.26). This has been depicted in the diagram shown in Fig. 2.2.

Among these, the Schrödinger equation and the path integral are concerned with distribution amplitudes. In particular, the real-valued path integral describes the spatiotemporal evolution of distribution amplitudes by summing over all possible paths with appropriate weights; however, the individual mathematical paths it contains are not dynamically determined. Both the Schrödinger function and the Green function represent coherent sums over infinitely many possible events.

In contrast, the quantum stochastic paths described by stochastic differential equations can be regarded as tracking the "physical fate" of individual events represented by each path in space-time. According to the stochastic differential equation for quantum paths, particles move under the influence of a velocity field coefficient that contains the Schrödinger vector as a functional, together with a driving Wiener process. While the velocity field coefficient mechanically drives trajectories, the Wiener process rejects the simple picture of a point particle with a definite velocity. However, this is not a wave, and wave should be understood not as an intrinsic property of the particle. Rather, it is a law or mode of motion associated with the Schrödinger function.

Therefore, it is appropriate to classify the three pillars into two categories: ("Schrödinger equation" + "path integral") and "quantum stochastic path dynamics." In this way, each of the three pillars possesses both strongly interrelated aspects and provide independent information and interpretation. This implies that one should not attempt to interpret all of quantum phenomena solely from the Schrödinger equation. In particular, the theory of quantum stochastic paths is still relatively new, and further intensive research is needed. We will focus on this aspect in the next chapter.

References

1. Nagasawa, M. Transformations of diffusion and Schrödinger processes. *Probability theory and Related Fields* **82**, 109–136 (1989).
2. Nagasawa, M. *Stochastic Processes in Quantum Physics*, vol. 94 (Birkhäuser, 2012).
3. Nagasawa, M. *Schrödinger Equations and Diffusion Theory* (Springer Science & Business Media, 2012).
4. Sanz, Á. S. & Miret-Artés, S. *A Trajectory Description of Quantum Processes. I and II. Applications: A Bohmian Perspective* (Springer, 2013).
5. Nelson, E. Derivation of the schrödinger equation from newtonian mechanics. *Physical Review* **150**, 1079 (1966).
6. Nelson, E. Review of stochastic mechanics. *Journal of Physics: Conference Series*, vol. 361, 012011 (IOP Publishing, 2012).
7. Yasue, K. Stochastic quantization: A review. *International Journal of Theoretical Physics* **18**, 861–913 (1979).
8. Yasue, K. Stochastic calculus of variations. *Journal of functional Analysis* **41**, 327–340 (1981).
9. Parisi, G. & Wu, Y.-s. Perturbation Theory Without Gauge Fixing. *Sci. Sin.* **24**, 483 (1981).
10. Namiki, M. *Stochastic Quantization*, vol. 9 (Springer Science & Business Media, 2008).
11. Takatsuka, K. On the dual structure of the schrödinger dynamics. *Journal of the Physical Society of Japan* **94**, 094003 (2025).
12. Newton, R. G. *Scattering theory of Waves and Particles* (Springer Science & Business Media, 2013).

The Nature and Role of Stochastic Paths in Quantum Mechanics

9

Abstract

This chapter highlights the nature and role of stochastic paths, with the subjects treated being;

1. Quantum nature seen from the classicalization of quantum paths
   1.1 Difference and similarity of quantum paths and Bohmian trajectories
   1.2 Quantum-mechanical canonical equations of motion and quantum Newton equation
   1.3 Reduction to classical dynamics; Hamiltonian canonical equations and Newtonian equation
2. Indirect correlations among individual quantum paths
   2.1 How does a part recognize the whole?
   2.2 The formation of interference (fringe) patterns in the double-slit experiment
3. The Wiener process as the essential feature of quantum path dynamics
   3.1 Time-reversal asymmetry
   3.2 Scaling law in Schrödinger dynamics through the path integral
   3.3 Scaling law and the Bohr model of the hydrogen atom: Avoidance of collapse of hydrogen atom
   3.4 Trajectories in a cloud chamber and orbits inside an atom: A question from Einstein to Heisenberg
4. Quantum stochastic processes and the uncertainty relation
   4.1 Position and momentum
   4.2 Time and kinetic energy
5. Quantum entanglement and spontaneous disentanglement on quantum paths
   5.1 Entanglement in the quantum path dynamics
   5.2 Spontaneous disentanglement (detanglement) in the path dynamics
   5.3 Disentanglement and Bell's inequality

K. Takatsuka, *Structure of the Schrödinger Dynamics*,
https://doi.org/10.1007/978-981-92-1910-0_9

Keywords

Roles of quantum stochasticity · Scaling property in path integrals · Dynamical roles of the Wiener process · Time irreversibility · No wavepacket collapse · Orbital concept in quantum mechanics · Dynamical origin of the uncertainty principle · EPR paradox · Spontaneous disentanglement · Bell's inequality

Here, in this chapter, we examine in detail the properties of quantum stochastic paths, which possess both the velocity field coefficient that works to deterministically drive the path and the Wiener process representing a random field that counteracts it. The essential difference of quantum path dynamics from classical mechanics lies in the fact that, in the former, both the Wiener process and $\bar{\psi}(q,t)$ appear in the equation of motion. We first focus on the velocity field coefficient and will discuss the Wiener process later.

9.1 Quantum Nature Seen from the Classicalization of Quantum Paths

We begin by examining the essential effects of quantum theory through the properties of the velocity field coefficient and the reduction of path dynamics to classical mechanics.

9.1.1 Quantum Paths and Bohmian Trajectories

In this book, Bohm's trajectory theory has already appeared several times. Sanz and Miret-Artés, in their book [1], succinctly describe each Bohmian trajectory as representing a streamline of the flow of $R(q,t)^2$ driven by S_B. The present author agrees with this view at the level up to the Schrödinger equation. However, in the context of quantum stochastic paths, it is reasonable to regard Bohmian trajectories as physical entities representing individual motions (provided that the Wiener process can be neglected). While I may seem to have argued in the former part of this book that they are merely one of possible representations of the Schrödinger function, in this chapter, we adopt a more advanced perspective.

Let us briefly review the Bohmian theory once again [2]. Bohm expresses the complex Schrödinger function using the real functions $R(q,t)$ and $S_B(q,t)$ as

$$\psi(q,t) = R(q,t)\exp\left(\frac{i}{\hbar}S_B(q,t)\right). \tag{9.1.1}$$

Here, S_B is supposed to satisfy an equation similar to the Hamilton-Jacobi (HJ) equation:

$$\frac{\partial S_B}{\partial t} + \frac{1}{2m}(\nabla S_B)^2 + V - \frac{\hbar^2}{2m}\frac{\nabla^2 R}{R} = 0, \tag{9.1.2}$$

the last term on the left-hand side of which is called the quantum potential. Regarding $R(q,t)$, $\rho(q,t) \equiv R(q,t)^2$ is subject to the equation of continuity:

$$\frac{\partial}{\partial t}\rho(q,t) = -\vec{\nabla} \cdot \vec{j}_B(q,t), \tag{9.1.3}$$

where the flux is given by

$$\vec{j}_B(q,t) = \frac{1}{m}(\nabla S_B)\,\rho(q,t) = v_B(q,t)\rho(q,t). \tag{9.1.4}$$

From the analogy with classical theory, it is immediately clear that $\nabla S_B/m$ corresponds to the quantum mechanical velocity $v_B(q,t)$. To highlight this feature, we define

$$v_B = \frac{1}{m}\nabla S_B = \frac{\hbar}{m}\,\mathrm{Im}\,\frac{\nabla\psi}{\psi}, \tag{9.1.5}$$

where ψ is the complex Schrödinger function, and by comparing this with Eq. (5.2.15), we see

$$\vec{p}^{\,\mathrm{local}}(q,t) = \vec{\nabla} S_B \tag{9.1.6}$$

as in [3]. Furthermore, by using Eqs. (8.2.22) and (5.2.19), it turns out that

$$\vec{\nabla} S_B/m = \vec{p}^{\,\mathrm{local}}(q,t)/m = \frac{\hbar}{m}\vec{\nabla}\theta(q,t) = \alpha^{\mathrm{Real}}(X_t,t). \tag{9.1.7}$$

Thus, it becomes clear that the velocity field in the stochastic differential equation (8.2.26) and the velocity field in the Bohm representation are equivalent. For Bohm, the physical meaning and origin of the quantum potential in Eq. (9.1.2) were of primary concern. However, from a different viewpoint, if the Schrödinger function is assumed to be given in advance and the Wiener process is absent from the stochastic differential equation (8.2.26), then the quantum stochastic paths coincide with Bohmian trajectories. Furthermore, the quantum stochastic paths may be viewed to proceed while stochastically wandering through the extended bundle of Bohmian trajectories with a fixed diffusion constant due to the presence of the Wiener process.

We note tha from the relation (9.1.7), it follows that Bohmian trajectories satisfy a non-crossing rule: If the Schrödinger function is given beforehand, the curves formed by two Bohmian trajectories do not intersect in (q,t)-space. This is because, if two trajectory curves starting simultaneously from the same Schrödinger function as an initial condition meet at some point, they are subsequently driven by exactly the same local momentum $\vec{p}^{\rm local}(q,t)$ and thus merge into a single trajectory. Therefore the Bohmian dynamics is well posed. This non-crossing rule corresponds to the fact that trajectory solutions of the Liouville equation do not intersect in the extended phase space.

Numerical examples of Bohmian trajectories have been increasingly reported in recent years. In particular, calculations of Bohmian trajectories in models of the double-slit experiment are well known and significant [1, 4]. Many numerical results show that Bohmian trajectories reproduce the fringe patterns appearing in the Schrödinger function while satisfying the non-crossing rule. In light of Eq. (9.1.7), this clearly indicates that interference patterns also emerge in the ensemble of quantum stochastic paths, provided that the randomness due to the Wiener process does not wash them out. It is also understood that a quantum stochastic path cannot pass through two slits simultaneously, and two Bohmian paths do not come to a single point on the detection panel at a time.

Finally, regarding Bohmian trajectories, it is worth mentioning the important work by Bob Wyatt [1, 5]. He proposed an algorithm that directly integrates Bohmian trajectories without evolving the Schrödinger function in space-time. Consider molecular dynamics, for example, chemical reactions. Since atomic nuclei are four to five orders of magnitude heavier than electrons, it is well known that, as a first approximation (the Born-Oppenheimer approximation), nuclei move relatively slowly along trajectories within the mean field generated by rapidly moving electrons. Wyatt and his collaborators aimed to extract nuclear motion directly as Bohmian trajectories and developed relevant effective computational methods. The key is how to handle the quantum potential $-\hbar^2/2m(\nabla^2 R/R)$ with use of only the concept of quantum trajectories. Although this is a highly challenging methodological approach, it is beyond the scope of this book, and we merely note its significance here [1, 5].

9.1.2 Stochastic Paths and Path Integrals

As aforementioned, if the Schrödinger function is given in advance, the curves formed by two Bohmian trajectories do not intersect in (q,t)-space. The stochastic paths can be regarded as a Wiener process defined on the manifold formed by those non-crossing trajectories. This non-crossing property may significantly constrain the summation over trajectories in the path integral, suggesting the possibility of employing quantum stochastic paths in path integral representations. It may be computationally more demanding to calculate quantum trajectories than to sum over random paths, and thus, the numerical advantage is not immediately clear. However, by summing a highly restricted set of trajectories according to the rules of Eqs. (6.2.7) and (6.2.8), one may obtain compact results in which the roles and contributions of individual trajectories become transparent. Since such calculations have not been carried out before, they are worth attempting.

9.1.3 Quantum Mechanical Canonical Equations of Motion and Quantum Newton Equation

It has been shown that the velocity field coefficient in the quantum stochastic differential equation is essentially identical to Bohm's quantum velocity. However, merely stating that "a trajectory is driven once the velocity field is given" does not reveal the structure of the dynamics. Therefore, let us transform Eq. (8.2.26) into an equation of motion.

9.1.3.1 Quantum Mechanical Canonical Equations of Motion (I)

First, in Eq. (8.2.26), we define the local momentum not normalized as

$$P_{X_t} = \left(\phi_r\ \phi_c\right) \hat{p} \left.\begin{pmatrix} \phi_r \\ \phi_c \end{pmatrix}\right|_{q=X_t}. \tag{9.1.8}$$

Heisenberg dynamics for this P_{X_t}, based on Eq. (5.2.1), yields

$$\begin{aligned}
\frac{d}{dt}\left(\phi_r\ \phi_c\right) \hat{p} \begin{pmatrix} \phi_r \\ \phi_c \end{pmatrix} &= \frac{1}{\hbar}\left(\phi_r\ \phi_c\right)\left[\hat{H}, \hat{p}\right] \mathbf{J} \begin{pmatrix} \phi_r \\ \phi_c \end{pmatrix} \\
&= -\left(\phi_r\ \phi_c\right)\left[V, \mathbf{J}\vec{\nabla}\right] \mathbf{J} \begin{pmatrix} \phi_r \\ \phi_c \end{pmatrix} \\
&= \left(\phi_r\ \phi_c\right)\left(-\vec{\nabla} V\right) \begin{pmatrix} \phi_r \\ \phi_c \end{pmatrix} \\
&= -\left(\vec{\nabla} V\right) \rho.
\end{aligned} \tag{9.1.9}$$

This equation is analogous to Ehrenfest's theorem, although, unlike Ehrenfest's theorem, it does not take the expectation value. Using Eq. (9.1.9) along with Eq. (8.2.26) together, we obtain a coupled ordinary differential equations as

$$\begin{cases} dX_t = \frac{1}{m\rho} P_{X_t} dt + \sqrt{\frac{\hbar}{m}} dW_0(t, \omega) \\ dP_{X_t} = -\left(\vec{\nabla} V(X_t, t)\right) \rho(X_t, t) dt, \end{cases} \tag{9.1.10}$$

which may be called the quantum mechanical canonical equations of motion.

In Eq. (9.1.10), the second equation (the equation for dP_{X_t}) shows that the force acting on the particles is modified by the distribution density function $\rho(X_t, t)$. In the first equation (the equation for dX_t), although the term $1/\rho(X_t, t)$ appears, its effect is not canceled out, and quantum effect appears to a certain extent through $\rho(X_t, t)$. In Bohm's theory, the quantum potential

$$-\frac{\hbar^2}{2m}\frac{\nabla^2 R}{R} \tag{9.1.11}$$

modifies the height of the original potential function $V(q)$, and it explains, for instance, that this may lower the barrier height affecting the tunneling probability. In Eq. (9.1.10), on the other hand, the "force" acting on the particle can be modified, thereby resulting in various quantum effects. The modified force $-\left(\vec{\nabla}V\right)\rho$ should be much easier to handle than the quantum potential. If an approximate algorithm can be devised to track ρ along the trajectory, as Wyatt aimed, quantum trajectories can be readily integrated. However, here, we should remember we have to take into account the Wiener process $W_0(t,\omega)$.

9.1.3.2 Quantum Newtonian Equation

In classical mechanics, it is possible to directly go from the Hamilton equations to the Newtonian equation. But in quantum theory, due to the presence of the Wiener process, there are some additional considerations that must be taken into account. First, by differentiating the first equation of (9.1.10) once more and substituting dP_{X_t} from the second equation

$$\begin{aligned} d^2X_t &= \frac{1}{m\rho}dP_{X_t}dt - \frac{P_{X_t}}{m\rho^2}d\rho dt + \sqrt{\frac{\hbar}{m}}d^2W_0(t,\omega) \\ &= -\frac{1}{m}\vec{\nabla}V(dt)^2 - \frac{1}{\rho}\vec{v}^{\text{local}}d\rho dt + \sqrt{\frac{\hbar}{m}}d^2W_0(t,\omega) \end{aligned} \tag{9.1.12}$$

is obtained. Furthermore, recalling Eqs. (5.1.30) and (5.2.15), we have

$$\begin{aligned} \frac{d\rho}{dt} &= -\vec{\nabla}\cdot\vec{j} = -\vec{\nabla}\cdot\frac{\hbar}{m}\left(\phi_r\vec{\nabla}\phi_c - \phi_c\vec{\nabla}\phi_r\right) \\ &= -\vec{\nabla}\cdot(\rho\vec{v}^{\text{local}}(q,t)), \end{aligned} \tag{9.1.13}$$

which reads in the style of the stochastic differential equation

$$d\rho = -\vec{\nabla}\cdot(\rho\vec{v}^{\text{local}}(q,t))dt. \tag{9.1.14}$$

Thus, putting this expression back into Eq. (9.1.12), we finally see

$$\begin{aligned} d^2X_t &= -\frac{1}{m}\vec{\nabla}V(dt)^2 - \frac{1}{\rho}\vec{v}^{\text{local}}d\rho dt + \sqrt{\frac{\hbar}{m}}d^2W_0(t,\omega) \\ &= -\frac{1}{m}\vec{\nabla}V(dt)^2 + \frac{1}{\rho}\vec{v}^{\text{local}}\left(\vec{\nabla}\cdot(\rho\vec{v}^{\text{local}})\right)(dt)^2 \\ &\quad + \sqrt{\frac{\hbar}{m}}d^2W_0(t,\omega). \end{aligned} \tag{9.1.15}$$

This is our quantum mechanical Newton equation. It takes a considerably different form from Nelson's stochastic Newton equation (8.1.10). Notice that the term proportional to $\vec{\nabla}\rho$ has appeared for the first in quantum dynamics only in the stage of the calculation of d^2X_t. Equation (9.1.15) closely resembles the original Newton equation, and it turns out that the terms after the first term on the right-hand side, including the Wiener process, have been augmented by quantum mechanics.

9.1.4 Quantum Mechanical Canonical Equations of Motion (II)

The coupled equations (9.1.10) have been transformed from the single differential equation (8.2.26), in which the velocity field is given as a function of the Schrödinger function, into a form that can be integrated once ρ on the right-hand side of the second equation is specified. The self-distribution dependence that appears through ρ (here, implying the contributions from other trajectories), together with the presence of the Wiener process, constitutes a characteristic feature of quantum mechanics. Therefore, we introduce the equation of continuity that determines ρ to Eq. (9.1.10) and regard them as a unified set of equations. From Eqs. (5.1.30) and (9.1.8), we have

$$\vec{j} = \frac{P_{X_t}}{m}. \tag{9.1.16}$$

Substituting this into the continuity equation (5.1.28), we obtain

$$\begin{aligned} \frac{d}{dt}\rho(q,t) &= -\nabla \cdot \vec{j}(q,t) \\ &= -\frac{1}{m}\nabla \cdot P_{X_t}, \end{aligned} \tag{9.1.17}$$

or formally

$$d\rho(X_t,t) = -\frac{1}{m}\nabla \cdot P_{X_t}dt \tag{9.1.18}$$

which is then added to Eq. (9.1.10). As a result, we obtain an apparently closed system of equations (vector notation is explicitly indicated) as

$$\begin{cases} d\vec{X}_t = \frac{1}{m\rho}\vec{P}_{X_t}dt + \sqrt{\frac{\hbar}{m}}dW_0(t,\omega) \\ d\vec{P}_{X_t} = -\left(\vec{\nabla}V(X_t,t)\right)\rho(X_t,t)dt \\ d\rho(X_t,t) = -\frac{1}{m}\vec{\nabla}\cdot\vec{P}_{X_t}dt. \end{cases} \tag{9.1.19}$$

Moreover, this system is composed entirely of classical quantities, so that the notion of the complementarity need not be considered; in other words, the phase term characteristic of quantum theory has disappeared. The structure of this system is such that one first integrates dP_{X_t} and $d\rho(X_t, t)$ simultaneously, then substitutes the resulting P_{X_t} and ρ into the equation for dX_t, and integrates it to obtain X_t under the Wiener process. The integration may be started from an initial point X_0 where $\rho(X_0, 0)$ is not negligible. Nevertheless, since the Wiener process and the term $\vec{\nabla} \cdot \vec{P}_{X_t}$ are involved, solving this system is not straightforward. It is also evident that the term $\vec{\nabla} \cdot \vec{P}_{X_t}$ reflects the self-distribution dependence of quantum dynamics.

As noted above, if the Wiener process is formally neglected in Eq. (9.1.19), a family of trajectories sharing a common Schrödinger function as the initial condition never intersect at the same point $\vec{X}_t$ at the same time. However, this differs from the situation in classical phase space, where multiple trajectories may pass through the same point q without intersecting if momenta are different. In quantum trajectories, for a given pair (X_t, t), there exists only a single momentum $\vec{P}_{X_t}$, determined through the velocity field coefficient. This formal momentum $\vec{P}_{X_t}$ is given, as in Eq. (9.1.17), by the mass m multiplied by the flux (see also (9.1.4)). This also reflects the role of $\rho(X_t, t)$ appearing in Eq. (9.1.19). The right-hand side of the second equation for $d\vec{P}_{X_t}$ indicates that the external force acting at the point X_t determines the increment of the flux (multiplied by mass) in proportion to the density at that point (see Eq. (9.1.16)), which has a meaning different from the classical equation

$$dp/dt = -\vec{\nabla} V(q, t). \tag{9.1.20}$$

We will later consider the reduction of the quantum canonical equations and quantum Newtonian equation to classical counterparts to clarify the above aspects of quantum dynamics.

$\rho(X_t, t)$ corresponds, in classical mechanics, to the density of trajectories in configuration space. The concept of trajectory density has long been studied in semiclassical mechanics and quantified through quantities such as the van Vleck determinant [6–9]. Alternatively, in terms of the phase-space distribution function obeying the Liouville equation discussed in Chap. 3.2, one may approximate the classical trajectory density function $\rho_{\text{classical}}(q, t)$ using the time-evolved distribution $\Gamma(q, p, t)$ as

$$\int dp \Gamma(q, p, t) \propto \rho_{\text{classical}}(q, t) \tag{9.1.21}$$

(where normalization should be treated separately). Such considerations provide insight into how $\rho(q, t)$ may be interpreted. However, since $\rho(q, t)$ in Eq. (9.1.19) represents the density of quantum trajectories, we do not pursue its relation to semiclassical theory here.

In the double-slit experiment shown in Fig. 2.1, the weight with which each individual trajectory reaches the observation point X_{obs} on the detection panel is given by $\rho(X_{\text{obs}}, t_{\text{obs}})$, where t_{obs} is the time required to reach the panel. Thus, the trajectories appearing through the velocity field coefficient in Eq. (8.2.26) may be regarded not merely as mathematical constructs representing the Schrödinger function, but as trajectories describing physical motion. However, the Wiener process obscures a causal and unique trajectory picture. The concept of trajectories, once denied by Heisenberg (to be discussed later in this chapter), thus partially re-emerges under these conditions. The interplay between trajectory concepts and stochastic processes will be discussed later in connection with the uncertainty relation and Bohr's hydrogen atom model.

9.1.5 Is Quantum Stochastic Process Dynamics a Law of Motion?

In Sect. 3.2 of Chap. 3, we outlined how Schrödinger dynamics corresponds to the laws of motion in classical mechanics. This was based on the viewpoint that the Schrödinger function is a spatially spread function obeying a partial differential equation, unlike classical particles governed by ordinary differential equations of ray solutions, and thus, the point of application of the acting "force" is ambiguous. Besides, in Chap. 5, the Schrödinger function was derived as a distribution amplitude function. From this perspective, just as the diffusion equation and the Fokker-Planck equation do not themselves represent laws of motion, the Schrödinger equation may also be regarded not as a law of motion but as a determining equation for a distribution function. The reason for distinguishing between these is the concern that a "law of motion" should explicitly describe both "cause" and "result (its effect)," whereas a "determining equation for a distribution function" appears to describe only the "result."

Nelson described his work as "Derivation of the Schrödinger equation from Newton mechanics" [10]. However, his stochastic Newton equation (8.1.10), although it appears to be a law of motion, is in fact a rule for a specific averaging procedure through $D\mathbf{X}_t$ in Eq. (8.1.1) and $D_*\mathbf{X}_t$ in Eq. (8.1.2). Similarly, in the Heisenberg equation and Ehrenfest's theorem, the Schrödinger equation acts through expectation values and does not represent local action on individual particles. On the other hand, quantum stochastic differential equations have the form of laws of motion, but since the velocity field coefficient is a function of the Schrödinger vector and is determined outside the equation itself, it is difficult to regard them as describing elementary dynamical processes. Even in the quantum canonical equations (9.1.19), although they are ordinary differential equations and the force $-\left(\vec{\nabla}V(X_t, t)\right)\rho(X_t, t)$ appears to act at X_t, the distribution nature of $\rho(X_t, t)$ prevents the elimination of the dynamical dependence on its own distribution.

Furthermore, due to the presence of the Wiener process in the first equation of Eq. (9.1.19), a stochastic indeterminacy independent of the force appears in dX_t. (While the Wiener process in the Brownian motion arises from direct interactions with the medium, no clear dynamical causality is presently identified for the quantum Wiener process.) Thus, even at the level of Eq. (9.1.19), quantum theory retains the essential feature that a point of action cannot be specified.

In the framework of Schrödinger dynamics, it may therefore be concluded that no uniquely determined causal law of motion based on localized interactions exists. Why is it that, even without knowledge of the elementary dynamical processes as "causes," we can nevertheless obtain the distribution rule (the Schrödinger equation) as the "effect," as well as the dynamics driven by it? Recall the discussion in Chap. 5: When we reconstructed the Schrödinger equation, the only conditions we imposed were the unitarity of the distribution amplitude function and the application of a variational principle to the equation of continuity. These are not discoveries of new physical interactions; rather, they are somewhat the "basic expectations" we bring to the theory.

Interestingly, despite these features, quantum stochastic differential equations reduce, in the limit $\hbar \to 0$, to the Hamilton canonical equations, or equivalently to the Newtonian equation, thereby directly recovering the classical laws of motion, as will be seen below.

9.1.6 Reduction to Classical Dynamics

Starting from the quantum canonical equations (9.1.10), let us consider the classical mechanical limit. This will make it easier to understand what fundamentally characterizes quantum theory.

9.1.6.1 Hamiltonian Canonical Equations

First, regarding the Wiener process in Eq. (9.1.10), since it is accompanied by a factor of $\sqrt{\hbar}$, the classical limit can be readily taken by letting $\hbar \to 0$. There is no problem in this limiting procedure.

In the quantum stochastic differential equation (8.2.26), the velocity field coefficient $\alpha^{\mathrm{Real}}(X_t, t)$ is given by Eq. (9.1.7) as $\alpha^{\mathrm{Real}}(X_t, t) = \nabla S_B/m$. In the limit $\hbar \to 0$, S_B becomes a solution of the classical Hamilton-Jacobi equation, and therefore, X_t coincides with the classical trajectory. However, here, we need to examine in somewhat more detail the physical situation in which X_t approaches a classical trajectory.

We now rewrite P_{X_t} in Eq. (9.1.8) in polar form, as in Eq. (5.2.17):

$$P_{X_t} = \hbar \rho \nabla \theta(X_t, t). \tag{9.1.22}$$

Note that we cannot simply take $\hbar \to 0$ and conclude that $P_{X_t} \to 0$, because $\nabla\theta(X_t, t)$ contains terms of order $\hbar^{-1}$. Specifically

$$\lim_{\hbar \to 0} \nabla\theta(X_t, t) \to \frac{p_{cl}(X_t, t)}{\hbar}, \tag{9.1.23}$$

where $p_{cl}(X_t, t)$ is the classical momentum.

Next, in the classical limit, the Wiener process disappears, so the trajectories no longer exhibit jumps or branching behavior but instead become continuous and smooth curves. Then, the density $\rho(X_t, t)$ carried along the trajectory should also remain equal to its initial value $\rho(X_0, 0)$. More precisely, in the Lagrangian view of fluid dynamics, for an incompressible fluid, along a streamline, one has $\rho(q, t) = \rho(q + v\Delta t, t + \Delta t)$, with v being the local velocity. Therefore, we may write

$$\rho(X_t, t) = \rho(X_0, 0) \equiv \rho_0, \qquad (9.1.24)$$

where ρ_0 is a constant specified along a given trajectory.

Accordingly, Eq. (9.1.22) becomes

$$P_{X_t} = \hbar \rho_0 \nabla\theta(X_t, t), \qquad (9.1.25)$$

which yields

$$P_{X_t} \to \rho_0 p_{cl}(X_t, t), \qquad (9.1.26)$$

and

$$dP_{X_t} \to \rho_0 dp_{cl}(X_t, t). \qquad (9.1.27)$$

Substituting this P_{X_t} into Eq. (9.1.10), we obtain

$$dX_t = \frac{1}{m} p_{cl}(X_t, t)dt, \qquad (9.1.28)$$

which is the familiar definition of classical momentum. Similarly, using Eq. (9.1.27), Eq. (9.1.10) can be rewritten as

$$dP_{X_t} = -\rho_0 \nabla V(X_t)\, dt. \qquad (9.1.29)$$

Combining Eqs. (9.1.27) and (9.1.29), we obtain

$$dp_{cl}(X_t, t) = -\nabla V(X_t), \qquad (9.1.30)$$

and thus the pair of Eqs. (9.1.28) and (9.1.30) reduces to the Hamiltonian canonical equations:

$$\begin{cases} \frac{dX_t}{dt} = \frac{1}{m} p_{cl}(X_t, t) \\ \frac{dp_{cl}(X_t, t)}{dt} = -\nabla V(X_t). \end{cases} \qquad (9.1.31)$$

As above, the key point has been the recovery of the classical force.

9.1.6.2 Quantum Newton Equation

The classical limit of the quantum Newton equation is again obtained by taking $\hbar \to 0$. First, the Wiener process disappears. As in Eq. (9.1.24), $\rho(X_t, t)$ may be regarded as constant along the trajectory, and therefore, it is expected that

$$d\rho(X_t, t) = 0. \tag{9.1.32}$$

As a result,

$$d^2 X_t = -\frac{1}{m} \nabla V \, (dt)^2, \tag{9.1.33}$$

which reproduces the classical Newtonian equation.

9.1.6.3 Relation Between Quantum and Classical Mechanics

From the above relation between quantum mechanics and classical mechanics, let us extract the conceptual implications.

First, let us confirm a natural fact: Classical mechanics is established without direct correlation between the part and the whole by erasing the information of $\rho(X_t, t)$ from the quantum equations of motion. In other words, classical trajectories as part of system dynamics can be calculated without the knowledge of $\rho(X_t, t)$ as the whole. Therefore, quantum theory can never be derived from classical mechanics. In this sense, the title of Nelson's paper [10], "Derivation of the Schrödinger equation from Newton mechanics," may be regarded as somewhat overstated. Conversely, it becomes clear what must be added to classical mechanics in order to restore quantum dynamics.

Here, we need to repeat the relations between quantum theory and classical mechanics as already listed in Sect. 3.2 in Chap. 3. In Bohm's formulation, the Schrödinger function is reduced to the quantum mechanical action $S_B(q, t)$ and the amplitude function $R(q, t)$ (Eqs. (9.1.2), (9.1.3)), and is related to the quantum Hamilton-Jacobi (HJ) equation. The HJ equation is a partial differential equation, and this corresponds to the fact that the Schrödinger function is a space-time distribution amplitude. The well-known WKB theory in semiclassical mechanics (more precisely, the lowest-order equation in the WKB expansion) [11, 12] also clearly expresses the relation between the Schrödinger equation and the original HJ equation without the quantum potential. Furthermore, it is well known that, in the $\hbar$-expansion of Wigner's phase-space distribution function [13–15], the lowest-order dynamics reduces to the classical Liouville equation. Thus, these correspondences concern the relation between quantum and classical theories at the level of distribution functions. On the other hand, the quantum stochastic path theory describes the dynamics of trajectories and, therefore, has a direct correspondence to the Hamiltonian canonical equations and/or Newtonian equation of motion.

This diversity of connections to classical mechanics reflects the dual nature of Schrödinger dynamics: the Schrödinger equation as a determining equation for the distribution amplitude function and, parallelly, quantum stochastic path dynamics as the dynamics of trajectories through the velocity field coefficient.

Reiterating, the essential difference between quantum paths and classical trajectories is that the density function $\rho(X_t, t)$ appears explicitly only in the quantum equations of motion, such as Eqs. (9.1.10) and (9.1.15). Conversely, to recover quantum behavior from classical mechanics, one must not treat $\rho(X_t, t)$ as constant along trajectories, but instead restore its reality and space-time extent. Besides, due to the Wiener process, quantum paths exhibit "jumps" in space, so that approximate values of $\rho(X_t, t)$ at the "landing points" are also required.

In the next-generation semiclassical theories and quantum technology as well, it will become both theoretically and practically important to incorporate the effects of the Wiener process effectively, in addition to the role of $\rho(q, t)$.

9.2 Indirect Correlations Among Individual Quantum Paths

9.2.1 Part and Whole: How Does a Part Recognize the Whole?

In the mechanical force term of the quantum canonical equations (9.1.10), $-\rho(X_t, t)\vec{\nabla}V(X_t, t)$ contains $\rho(X_t, t) = |\psi(q, t)|^2$. This implies self-distribution dependence in the sense that the acting "force" is not determined solely by the local circumstances along an individual trajectory but by other path distribution.

Strictly speaking, $\psi(q, t)$ is the sum of contributions from all trajectories and is determined by the entirety of Schrödinger dynamics (i.e., contributions from other trajectories). The global structure is determined by incorporating all trajectories into the path-integral representations, such as Eqs. (6.2.6) or (6.2.7). As a result, self-referential nonlinear correlations arise among the quantum paths. Each individual trajectory cannot exist independently in isolation, but instead evolves in an environment that appears as if nonlinear mutual interactions are present. In classical mechanics, such self-referential nonlinear relationships between trajectories completely disappear even though the non-crossing rule is applied to the paths in $\Gamma(q, p, t)$.

To be a little more specific, even in the quantum stochastic path dynamics, each path $X_t(\omega)$ is generated by solving the equations of motion of Eq. (9.1.19). Therefore, each path seems independent from others just as classical trajectories are. However, since the velocity field coefficient that drives the quantum paths is a functional of the Schrödinger function (the distribution amplitude function), each individual trajectory is driven based on the "knowledge" of the global distribution, while the global distribution density is the limit of the sum of contributions from individual trajectories.

A similar situation holds in diffusion theory. Recall Eq. (4.4.8), where ψ_D appeared as the solution of the diffusion equation. However, in diffusion phenomena, approximate descriptions such as the Langevin equation are available, so that individual Brownian paths can be computed without resorting to stochastic differential equations. Regarding the relation between the global distribution and individual paths in Brownian dynamics, the individual paths that ultimately construct the distribution function can be regarded as independent events (assuming that collisions among multiple Brownian particles are ignored). Paths occurring at different times are collected together and quantified as a mathematical distribution function. In other words, individual Brownian paths can be tracked even by the eye of a dog, whereas the quantum paths and the associated mathematical distribution function are an image that can only be projected in the human mind.

Let me repeat the above aspect to emphasize the critical aspect of Schrödinger dynamics: Individual paths possess "mutual indirect correlations" through the distribution amplitude function underlying the velocity field coefficient in Eq. (8.2.26). Even though multiple paths do not directly influence each other through interactions such as Coulomb forces, the ensemble of paths as a whole is guided so as to reproduce the distribution density ρ appearing in the equations of motion (9.1.10) and (9.1.19) and thus behave as if they were mutually coordinated.

Therefore, in quantum dynamical systems where the Schrödinger function can change significantly only locally even under slight variations of experimental conditions, the geometric shapes of individual paths and their indirect correlations may change substantially. This should be regarded as a new issue in the so-called quantum chaos. In classical chaos studies, the degree of chaos can naturally be quantified based on the instability of trajectories under small variations of system parameters [16]. In contrast, in traditional quantum chaos theory, it has been customary to define the degree of chaos from correlation functions based on the time evolution of the Schrödinger function and anomalies in the associated spectral structures [17–21]. Thus, investigating dynamical instability in quantum systems from the standpoint of quantum stochastic paths and the quantum ergodicity as well have acquired new significance in the present era of advanced quantum technologies.

9.2.2 On the Formation of Interference (Fringe) Patterns in the Double-Slit Experiment

Based on the above discussion, let us consider the double-slit experiment demonstrated in Chap. 2 from the viewpoint of quantum stochastic paths.

Since the Schrödinger function is a distribution amplitude function and obeys a linear equation, it is natural to assume that the Schrödinger function passing through two slits can be expressed as the simple sum of the distribution amplitude functions corresponding to each slit individually. This is not surprising, since the distribution amplitude function is a mathematical object visible only to the human mind. However, it is necessary to properly take into account the phases of the two components and to assume that components reflected from the slit plate without passing through the slits do not interfere with the transmitted components.

To reiterate, the fact that the Schrödinger function passes through two slits simultaneously and that a linear superposition can be taken is a property of the distribution amplitude function and is distinct from the dynamics of individual electrons. The dynamics of individual electrons must be described by quantum stochastic path dynamics. Statements often found in books, such as "an electron passes through two slits simultaneously and interferes with itself" or "it has no partner to interfere with other than itself," are simply incorrect. In quantum stochastic path dynamics, it is clear that each individual path cannot bifurcate and pass through both slits.

Experimentally, as seen in the right panel of Fig. 2.1, the spots formed by electrons arriving at the detection screen initially appear random. However, as the number of observed particles increases, a pattern gradually emerges, eventually forming the so-called fringe pattern. As discussed above, individual quantum stochastic paths appear to evolve independently, but as an ensemble, they are guided so as to reproduce $\psi(q,t)$ given by the Schrödinger equation. (Recall also that the Schrödinger equation works as a machinery to make the equation of continuity stationary as in Eq. (5.1.40).) Consequently, the observed distribution converges to the (robust) distribution $|\psi(q,t)|^2$. Importantly, the magnitude of $|\psi(q,t)|$, which makes the direct contribution to the formation of the fringe pattern, emerges through a mechanism independent of the concept of "interference" from the viewpoint of quantum stochastic path dynamics.

As repeatedly emphasized, quantum stochastic paths coincide with Bohmian trajectories if the Wiener process is ignored. In double-slit models, Bohmian trajectories have been extensively computed [1, 4], and they successfully reproduce the fringe pattern. For example, see the Bohmian trajectories shown in Fig. 9.1 taken from Philippidis et al. [4]. These trajectories are guided toward the directions where fringe patterns appear. This is not because a single electron interferes with itself, but rather due to the non-crossing rule of Bohmian trajectories. That is, given the Schrödinger function, Bohmian trajectories do not cross in (q,t)-space. This non-crossing rule is clearly reproduced in Fig. 9.1 and is one of the manifestations of the indirect correlation among the quantum paths.

Because of this rule, it is impossible for two trajectories to overlap at the same space-time point and interfere with each other. Although Bohmian trajectories carry the quantum phase $S(q,t)/\hbar$, the non-crossing property implies that, through the relation $|\psi(q,t)| = R(q,t)$, the magnitude of $|\psi(q,t)|$ is the reflection of the initial distribution $|\psi(q,0)|$ and the density of trajectories. Thus, by tracking trajectory density, one can determine the spatial distribution structure such as the fringe pattern of $|\psi(q,t)|$ or $|\psi(q,t)|^2$ at the observation point.

For the same reason, in regions where trajectories from the two slits appear to merge (central region), the trajectories do not intersect but are arranged in a geometrically parallel manner. If only one slit were present, the distribution would exhibit symmetry centered at the slit position on the detection screen, provided that the electron gun is set normal to the panel and slit. From the viewpoint of a single slit, therefore, local symmetry is clearly broken at the merging region, leading naturally to an increase in trajectory density.

A closer inspection of Fig. 9.1 shows that the point of avoided confluence occurs in upstream regions of high trajectory density, and the distortion of flow develops downstream while being amplified. These density fluctuations eventually produce alternating dense and sparse regions downstream, forming the fringe pattern as a result.

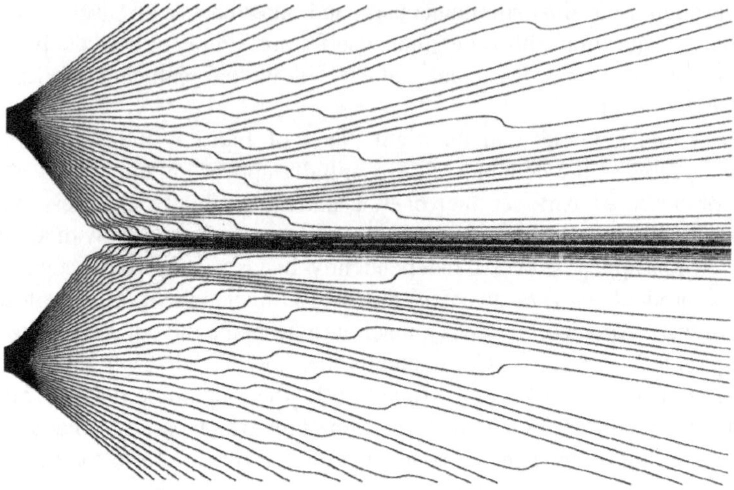

Fig. 9.1 Bohmian trajectories in the double-slit experiment. Only the trajectories after passing through the slits are shown. Reproduced from the seminal work by Philippidis et al. [4] with permission

However, quantum stochastic paths are continuous trajectories undergoing Wiener process along Bohmian trajectories. If the stochastic effect is strong, the fringe pattern tends to be smoothed accordingly. Furthermore, as seen in Fig. 2.1, the apparent randomness of individual spots arises both from the impossibility of perfectly controlling initial conditions due to the uncertainty principle and from intrinsic fluctuations induced by the Wiener process. Moreover, as will be discussed later, the Wiener process itself is one of the origins of the uncertainty relation, and the two cannot be separated in principle.

From this viewpoint, it is not mathematically mysterious that a fringe pattern is formed even when electrons are emitted one by one. The real mystery lies in how an individual electron can "sense" the global distribution $\rho = |\psi(q,t)|^2$, namely, the functional relationship between part and whole.

To examine this question, we return to Eq. (5.2.19). The local momentum $\vec{p}^{\text{local}}(q,t)$, the velocity field coefficient $\alpha^{\text{Real}}(X_t,t)$ in Eq. (8.2.26), and the local phase $\theta(q,t)$ of the Schrödinger vector are related by Eq. (9.1.7), as displayed here again as

$$\frac{1}{m}p^{\text{local}}(X_t,t) = \alpha^{\text{Real}}(X_t,t) = \frac{\hbar}{m}\vec{\nabla}\theta(q,t). \tag{9.2.1}$$

Thus, $\theta(q,t)$ acts as the driving factor of particle motion (excluding the Wiener process). Given the initial and boundary conditions, one may regard that a "field" of the phase $\theta(q,t)$ is generated as a distribution amplitude function. In classical diffusion as well, Brownian particles in the Ito stochastic differential equation may also be regarded as moving on a field generated by the function subject to the diffusion equation. However, since both the Schrödinger vector and the solution of the diffusion equation are mathematical representations of distributions rather than physical entities, it is difficult to explain how the θ-field can possess physical reality.

Let us now consider another aspect of $\vec{p}^{\text{local}}(q,t)$. Comparing Eq. (5.2.15)

$$\vec{p}^{\text{local}}(q,t) = \frac{\hbar}{\rho}\left(\phi_r\vec{\nabla}\phi_c - \phi_c\vec{\nabla}\phi_r\right) \tag{9.2.2}$$

with the flux expression (5.1.30)

$$\begin{aligned}\vec{j} &= \left(\phi_r(q,t)\ \phi_c(q,t)\right)\frac{\hat{p}}{m}\begin{pmatrix}\phi_r(q,t)\\ \phi_c(q,t)\end{pmatrix}\\ &= \frac{\hbar}{m}\left(\phi_r\vec{\nabla}\phi_c - \phi_c\vec{\nabla}\phi_r\right)\end{aligned} \tag{9.2.3}$$

we obtain

$$\frac{1}{m}\vec{p}^{\text{local}}(q,t) = \frac{1}{\rho(q,t)}\vec{j}(q,t).$$

On the other hand, from Eq. (9.2.1)

$$\begin{aligned}\alpha^{\text{Real}}(X_t,t) &= \frac{1}{m}\vec{p}^{\text{local}}(X_t,t)\\ &= \frac{1}{\rho(X_t,t)}\vec{j}(X_t,t).\end{aligned} \tag{9.2.4}$$

Then, neglecting the Wiener process in Eq. (8.2.26), we have

$$dX_t = \alpha^{\text{Real}}(X_t,t)dt, \tag{9.2.5}$$

which can be rewritten as

$$\rho(X_t,t)dX_t = \vec{j}(X_t,t)dt. \tag{9.2.6}$$

In Bohmian theory, as written in Eq. (9.1.4), it holds that

$$\vec{j}_B(q,t) = \frac{1}{m}\left(\vec{\nabla}S_B\right)\rho(q,t) = \vec{v}_B(q,t)\rho(q,t) \tag{9.2.7}$$

and therefore Eq. (9.2.6) becomes

$$dX_t = \vec{v}_B(q,t)dt = \frac{1}{m}\left(\vec{\nabla}S_B\right)dt. \tag{9.2.8}$$

This supports a view that Bohmian trajectories may be regarded as physical particle trajectories (again neglecting the Wiener process).

If, instead of Eq. (9.2.7), one assumes the classical fluid flux

$$\vec{j}(q,t) = \vec{v}^{\text{classic}}(q,t)\rho(q,t) \tag{9.2.9}$$

then $\rho(q,t)$ cancels in Eq. (9.2.6), and the "whole" governing the "part" disappears, merely yielding

$$dX_t = \vec{v}^{\text{classic}}(X_t,t)dt, \tag{9.2.10}$$

which is the familiar equation of classical mechanics.

In summary, although the specific form of $\vec{j}(X_t,t)$ of Eq. (9.2.3) in quantum theory produces an essential deviation from classical mechanics, the combination of Eqs. (9.2.5) and (9.2.6) is not far from the classical counterpart. After all, apart from the Wiener process, X_t is ultimately driven so as to reproduce the flow generated by $\rho(q,t)$.

9.2.3 Is the Schrödinger Function a Pilot Wave?

Bohm trajectories are said to have been conceived as the motion of particles guided by de Broglie's idea of the pilot wave. However, if one examines Eq. (9.2.1), it appears more natural to regard the quantum motion as being driven by the gradient of the rotational field generated by the Schrödinger vector, namely, $\vec{\nabla}\theta(q,t)$.

Even so, since the quantum stochastic dynamics requires the Schrödinger function to construct the "velocity field" beforehand, the pilot-wave interpretation may well give an impression that is not entirely off the mark.

However, in Chap. 5, we have shown that the Schrödinger equation possesses the function of selecting the most physically probable (or robust) Schrödinger vector $\bar{\psi}(q,t)$ within the continuity equation (5.1.39) for $\rho(q,t)$ through the variational principle, Eq. (5.1.40). That is, the Schrödinger equation selects the most robust set of $\rho(q,t)$ and $j(q,t)$ under given conditions. Therefore, one may add the following statement: "X_t undergoes random motion due to the Wiener process along the streamlines corresponding to the most robust flux generated from the given initial conditions."

In the case of the double-slit experiment, each individual quantum stochastic trajectory appears so as to "reproduce" the most robust (most probable in physical sense) $\rho(q,t)$ and $j(q,t)$ under the given experimental conditions. Each such trajectory corresponds to a single point in the right panel of Fig. 2.1, and the ensemble ultimately converges to $\bar{\psi}(q,t)^T\bar{\psi}(q,t)$.

Accordingly, de Broglie's idea had better be reinterpreted as follows. Each trajectory is not guided by a pilot wave; rather, it is selected so as to reproduce the robust $\rho(q,t)$ and $j(q,t)$ determined by the variational principle of Eq. (5.1.39). In fact, the quantum canonical equations (9.1.10) and (9.1.19) describe how the position $\vec{X}_t$, the momentum $\vec{P}_{X_t}$, and the spatial density of trajectories $\rho(X_t, t)$ are mutually correlated at each instant and jointly drive the motion. Thus, the dynamics are determined without invoking the notion of a "guiding wave."

9.3 The Wiener Process as the Essential Feature of Quantum Path Dynamics

Let us retrace the classicalization procedure carried out in Sect. 9.1.6 of this chapter. In that procedure, (i) the Wiener process disappears as $\hbar \to 0$; (ii) as a result, the trajectory becomes deterministic and uniquely traceable without branching or jumps; and (iii) consequently, $\rho(X_t, t) = \rho(X_0, 0) \equiv \rho_0$ and $d\rho(X_t, t) = 0$ are expected, ultimately leading to the Hamilton canonical equations and Newton's equations.

Thus, within the framework of quantum path dynamics, "if no stochastic process is associated with the path, the dynamics can be reduced to classical deterministic mechanics." Conversely, "if the path does not reduce to classical deterministic mechanics, it is because a stochastic process is associated with the path." Alternatively, even if the velocity field coefficient has a quantum-mechanical form, without the stochastic process, it merely reduces to Bohm trajectories and the Schrödinger equation and does not represent quantum paths corresponding to individual events.

Therefore, we next discuss the reality of the Wiener process contained in Eq. (8.2.26) and its effects on quantum dynamics.

Recall that Nagasawa, based on the theories of Kolmogorov and Ito and through his own rigorous mathematical developments, clarified that stochastic processes necessarily exist in quantum path dynamics [22, 23]. If a quantum particle obeys the Schrödinger equation, then a stochastic process inevitably underlies it. We have also discussed in this book the structure of quantum path dynamics and its velocity field coefficients while explicitly showing the close relationship and duality between diffusion theory (or stochastic process theory) and Schrödinger dynamics. That is, the velocity field coefficient represents precisely the velocity field generated by the Schrödinger function. Moreover, the diffusion coefficient of the Wiener process that appears, $D = \hbar/2m$, is exactly the "quantum diffusion coefficient" appearing in the Schrödinger equation (6.2.2). More intuitively, this can be regarded as one of the consequences of the fact that the first-order time derivative and the second-order spatial derivative are directly linked within a single equation (the Schrödinger equation) (see Eq. (6.2.2)).

However, the physical origin of stochastic processes in classical diffusion phenomena and in quantum dynamics is completely different. To be honest, the author cannot attribute the origin of the Wiener process in quantum stochastic processes to any other physical phenomenon. At present, it is regarded merely as a property related to the infinitesimal limit of space-time.

In the following, we discuss quantum effects, phenomena, and pictures that emerge from stochastic processes and cannot be clarified by the Schrödinger equation alone.

9.3.1 Time-Reversal Asymmetry

We have repeatedly emphasized in this book that the Schrödinger equation and the diffusion equation differ in their time evolution in that the former possesses unitarity, while the latter has a semigroup property. In their stochastic differential equation representations, both are expressed in a form that is irreversible under time reversal. In quantum theory, these are Eqs. (8.2.26) and (9.1.10) and, in diffusion theory, Eq. (4.4.8). (However, this does not mean that the general stochastic differential equation theory is always time-reversal asymmetric.)

On the other hand, as a fundamental property of stochastic processes, the Wiener process satisfies

$$\left\langle \frac{dW_0}{dt} \right\rangle = 0 \tag{9.3.1}$$

and for the mean trajectory

$$\langle dX_t \rangle = \left\langle -\frac{1}{\rho} \begin{pmatrix} \phi_r & \phi_c \end{pmatrix} \frac{\hat{p}}{m} \begin{pmatrix} \phi_r \\ \phi_c \end{pmatrix} \right\rangle dt. \tag{9.3.2}$$

Thus, the time-reversal symmetry of the average trajectory differs from that of each individual trajectory.

Similarly, the Schrödinger equation contains no Wiener process and is known to possess time-reversal symmetry. In this way, the Schrödinger equation has its own intrinsic symmetry, which may differ from the properties of individual particles (events), and this must be carefully noted in understanding quantum phenomena. Conversely, it may not always be possible to extract asymmetric individual motions from a time-reversal symmetric Schrödinger equation, since some information—such as that represented by Eq. (9.3.1)—may cancel out upon averaging.

There may be other such quantities as well. The problem of "strange interactions in quantum entanglement and spontaneous disentanglement," to be discussed later, is also related to this issue.

9.3.2 Scaling Law of Schrödinger Dynamics

The fundamental property of the stochastic process underlying quantum theory

$$(dW)^2 = (\hbar/m)\, dt \tag{9.3.3}$$

is of the same nature as the scaling law in Brownian motion

$$\left\langle \Delta q^2 \right\rangle = 2D\Delta t. \tag{9.3.4}$$

Incidentally, Eq. (9.3.4) is called a scaling law because Mandelbrot generalized Brownian motion into the form

$$\left\langle \Delta q^2 \right\rangle = 2D\Delta t^{2H}, \tag{9.3.5}$$

(called fractional Brownian motion) [24] and positioned it as one of the geometrical and physical phenomena that obey general scaling rules. Following this convention, we shall also refer to the relation between Δq^2 and Δt as a scaling rule in the present discussion of quantum theory.

As is well known, this scaling law imposes restrictions on the limiting operations $\Delta q \to 0$ and $\Delta t \to 0$. For example, the derivative

$$\lim_{\Delta t \to 0,\, \Delta q \to 0} \frac{\Delta q}{\Delta t} \tag{9.3.6}$$

and $\Delta q \to 0$ are incompatible. Therefore, the function $q = q(t)$ expressed as a function of t may lose smoothness (i.e., it may not be continuously differentiable). Classical examples of functions that are nowhere differentiable include the Weierstrass function and the Takagi function. The theories of stochastic processes by Ito and Stratonovich [25], as well as Mandelbrot's fractal geometry [24], play important roles in mathematics and the natural sciences.

The scaling law of stochastic processes, Eq. (9.3.3), corresponds to the scaling of the skew Wiener measure in the real quantum path integral, Eq. (6.2.8). It is given by

$$d\mathbf{P}_{W[q,t:0,0]}(\omega) = \lim_{N\to\infty} \left(\frac{m}{2\pi\hbar \mathbf{J}\Delta s} \right)^{(N-1)/2}$$
$$\times \prod_{k=1}^{N-1} dq_k \exp\left[\Delta s \left(-\frac{m}{2\hbar} \left(\frac{\Delta q_k}{\Delta s} \right)^2 \right) \mathbf{J}^T \right]. \tag{9.3.7}$$

The exponent is dimensionless and can be normalized in the form of a standard deviation as

$$\left\langle \frac{m}{2\hbar} \left(\frac{\Delta q_k}{\Delta s} \right)^2 \Delta s \right\rangle = \frac{1}{2}. \tag{9.3.8}$$

This leads to

$$\left\langle \frac{\Delta q^2}{\Delta t} \right\rangle = \frac{\hbar}{m}, \tag{9.3.9}$$

which coincides with the scaling law of Brownian motion, Eq. (9.3.4), with $D = \hbar/2m$. Of course, in Feynman's path integral, Eq. (6.1.6), if the kinematic part is normalized as

$$\left\langle \frac{m}{2\hbar} \frac{\Delta q^2}{\Delta t} i \right\rangle = \frac{1}{2} i, \tag{9.3.10}$$

we likewise obtain

$$\left\langle \Delta q^2 \right\rangle = \hbar / m \Delta t \tag{9.3.11}$$

as expected. Here, in the path integral formula, we have taken $\left\langle (q_{k+1} - q_k)^2 \right\rangle = \left\langle \Delta q^2 \right\rangle$.

Equation (9.3.9) admits various simple transformations. If Δt is regarded as an externally given parameter, one may consider that the system is confined within a region of size approximately $m \Delta q^2 / \hbar$ (a "surface" in one dimension) in average. However, the exact position within this region is unknown. Writing Eq. (9.3.9) as

$$\left\langle \Delta q \times \left(m \frac{\Delta q}{\Delta t} \right) \right\rangle = \hbar, \tag{9.3.12}$$

we see, as discussed above

$$\lim_{\Delta t \to 0, \Delta q \to 0} \left\langle m \frac{\Delta q}{\Delta t} \right\rangle = 0, \tag{9.3.13}$$

that is, $\Delta p \to 0$ and $\Delta q \to 0$ cannot be simultaneously realized. Therefore, one must abandon the possibility that $\langle \Delta q \rangle = 0$ and $\langle \Delta p \rangle = 0$ hold simultaneously. Of course, this is related to the uncertainty relation, which will be discussed later. In this way, although there is no medium in quantum dynamics that randomly kicks in contrast to Brownian motion, a scaling law exists in quantum theory that restricts the average motion. We now examine its effects.

9.3.3 Scaling Law and the Bohr Model of the Hydrogen Atom: Avoidance of Atomic Collapse

Next, we demonstrate that the scaling law, Eq. (9.3.9), can have actual physical consequences. As an example, we consider the Bohr model of hydrogen-like atoms, which played a crucial role in the birth of quantum mechanics. Let us consider a one-dimensional hydrogen atom. An electron with charge $-e$ moves back and forth along a straight line passing through a nucleus with charge $+Ze$. If the electron is, on average, located at a distance Δq from the nucleus, the classical Hamiltonian, written in finite-difference form, is

$$H = \frac{1}{2} m \left(\frac{\Delta q}{\Delta t} \right)^2 - \frac{Ze^2}{\Delta q} \tag{9.3.14}$$

In what follows, all equations are understood in the sense of averages. Of course, $\Delta q/\Delta t$ corresponds, in the limit, to the average relative velocity of the electron. Rather than integrating this equation to obtain trajectories, we explicitly incorporate the constraint given by Eq. (9.3.9) as

$$\frac{\Delta q}{\Delta t} = \frac{\hbar}{m}\frac{1}{\Delta q} \tag{9.3.15}$$

and substitute it into Eq. (9.3.14) to reduce the energy to

$$H = \frac{1}{2}\frac{\hbar^2}{m}\Delta q^{-2} - Ze^2\Delta q^{-1}. \tag{9.3.16}$$

Then

$$\frac{dH}{d(\Delta q)} = -\frac{\hbar^2}{m}\Delta q^{-3} + Ze^2\Delta q^{-2} = 0 \tag{9.3.17}$$

gives the points where the forces are balanced or where the energy takes extremal values. These are found at $|\Delta q| = \infty$ and

$$\Delta q = \frac{\hbar^2}{Zme^2} \equiv \frac{a_0}{Z}. \tag{9.3.18}$$

Here, a_0 is the celebrated Bohr radius (~0.529 Å) defined in Eq. (9.3.18), which corresponds to the radius of the 1s orbital of the hydrogen atom in the Bohr model [12]. In this one-dimensional model, there is no circular orbit; rather, the electron should be considered to be located at a distance a_0/Z from the nucleus in average. The energy at this point is obtained by substituting Eq. (9.3.18) into Eq. (9.3.14) as

$$E = -\frac{1}{2}\frac{Z^2me^4}{\hbar^2}, \tag{9.3.19}$$

which exactly agrees with the ground-state energy in the Bohr model [12].

At the beginning of Chap. 8, when introducing Nelson's theory, we quoted from the introduction of his paper [10]: "If there were a hydrogen atom in the ground state, the electron would be in a dynamic equilibrium between the attractive force of the nucleus and the random force due to Brownian motion. ⋯." This picture is somewhat peculiar, and its basis is not immediately clear. However, as seen above, if it refers to the role of the Wiener process directly reflected in the eigenenergy of the hydrogen atom, one may gain some understanding of what Nelson might intend to convey.

Returning to the scaling of Feynman's path integral, Eq. (9.3.10), if we allow scaling with respect to harmonics, the condition is relaxed for natural numbers n as

$$\left\langle \frac{m}{2\hbar} \frac{\Delta q^2}{\Delta t} \right\rangle i = \frac{1}{2} n i. \tag{9.3.20}$$

This is equivalent to

$$\left\langle m(\frac{\Delta q}{\Delta t})\Delta q \right\rangle = n\hbar, \tag{9.3.21}$$

which, by interpreting $v = \Delta q/\Delta t$ (velocity) and $\Delta q = r$ (radius), becomes identical to the quantum condition for the angular momentum in Bohr orbits [12]

$$mvr = n\hbar. \tag{9.3.22}$$

Accordingly, in Eqs. (9.3.18) and (9.3.19), replacing $\hbar$ with $n\hbar$ reproduces the excited states of the Bohr model.

Why does the electron not emit electromagnetic radiation and fall into the nucleus? In the Bohr model, this is explained by noting that, from condition (9.3.22), when $n = 0$, one obtains $\Delta q = 0$ and $E = -\infty$, so that such a state does not exist. Does this truly constitute an explanation? In contrast, under the scaling law, if the electron were to fall to $\Delta q = 0$, Eq. (9.3.18) would imply that $|\Delta q/\Delta t|$ diverges to infinity, thereby preventing $\Delta q = 0$, and consequently, the system cannot satisfy the stability condition (9.3.17). This is essentially the same as an explanation based on the uncertainty relation.

Some readers may recall, from their first course in quantum mechanics long ago, diagrams of the hydrogen atom in early quantum theory, where an electron moves along a circular orbit around the nucleus, and a wave—intended to represent a "matter wave"—is superimposed along the circumference. One then checks whether an integer multiple of the wavelength matches the length of the circumference. What, then, was this "wave-like object" that was presumably used as an illustration of a matter wave? Many freshmen may have been led to believe that it was a physically real wave. Even today, this way of introducing the concept of quantum condition persists in classrooms, often accompanied by a certain (slight?) guilty conscience.

Incidentally, the idea of quantization of the electron's angular momentum introduced by Bohr was extended by Sommerfeld to the quantization of action variables. Subsequently, the theory of quantization for integrable systems was completed in the form of the Einstein-Brillouin-Keller (EBK) quantization condition [8, 12, 26]. Keller contributed the 1/2 term in the harmonic oscillator expression

$$\frac{1}{2\pi} \oint p dq = (n + 1/2)\hbar \qquad n = 0, 1, 2, \cdots . \tag{9.3.23}$$

This quantity was further systematized using the Maslov index (the α_k in the next equation) and the Maslov phase as

$$\frac{1}{2\pi}\oint_{C_k}\sum_i p_i dq_i = (n_k + \frac{\alpha_k}{4})\hbar. \tag{9.3.24}$$

In this expression, C_k denotes the k-th independent cycle integral along a torus. Einstein published his work on the quantization of such integrable systems around 1917, during the same period in which he was actively engaged in the study of general relativity. In that paper, he clearly pointed out the limitations of his quantization theory, originating from the Bohr model, by noting that his formulation could not be applied to non-integrable systems. This is due to the absence of a torus to be quantized in phase space. The quantization of non-integrable systems later regained interest in the era of chaos beginning in the 1950s, and starting from Gutzwiller's periodic orbit theory [27,28], various interesting mathematical theories were developed [17–21, 27–29]. The study of chaos in Hamiltonian systems has deepened our understanding of the mathematical structures underlying classical and quantum mechanics, including their geometrical aspects.

Although there are a few physical problems that can be solved solely using scaling laws, the quantization of the harmonic oscillator can also be achieved, except for the Maslov index. These examples suggest that the scaling law indeed manifests itself as a physical effect. It should be emphasized that this scaling law does not arise from noise, nor is it merely an analogy to Brownian motion. Rather, it is a consequence of the stochastic differential equation (8.2.26) and the quantum path integrals (6.2.7) and (6.2.8). Although the quantum Wiener process is averaged out in the Schrödinger equation, this does not mean that it can always be neglected. In particular, it may play an important role when studying individual events.

9.3.4 Trajectories in a Cloud Chamber and Orbits Inside an Atom: A Question from Einstein to Heisenberg

With the theories developed by Heisenberg and Schrodinger, the program of quantizing Bohr's classical orbits became entirely a thing of the past, and the concept of "orbit" was no longer required to explain the spectrum of the hydrogen atom. Moreover, Heisenberg arrived at the view that "a physical theory must be constructed only from what are observable." Around that time, Einstein asked Heisenberg, "Although in a cloud chamber one can clearly see the trajectories of particles, you say that the concept of an (electronic) orbit inside an atom must be completely abandoned?" To this, Heisenberg is said to have replied, "Inside an atom, it is not possible to observe the orbit of an electron." (W. Heisenberg, *Der Teil und Das Ganze* or *Physics and Beyond*' [30].) However, from today's perspective, this response does not seem to fully address Einstein's question. As Einstein pointed out, at the scale of a cloud chamber (or more generally at macroscopic scales in classical mechanics), the concept of trajectory is clearly valid, yet the answer does not clarify under what circumstances the concept of trajectory becomes invalid. More positively, one would like to understand the relative magnitude between the velocity field coefficient (and their classical limit as $\hbar \to 0$) and the Wiener process that tends to wash them out.

Let us therefore evaluate $\hbar/m$ in the quantum stochastic process equation (9.3.3), $(dW)^2 = (\hbar/m)\, dt$. For an electron, this gives 1.158×10^{-2} Å^2 as^{-1}. Substituting this back into Eq. (8.2.26), we find that, at the order of magnitude $\sqrt{\hbar/m} = 1.076 \times 10^{-1}$ (Å as$^{-1/2}$), the trajectory driven by the velocity field coefficient competes with the quantum diffusion process. If we denote the spatial spread of the particle motion by ΔR and the corresponding time scale by ΔT, then when the magnitude of $\Delta R \Delta T^{-1/2}$ is comparable to or smaller than $\sqrt{\hbar/m}$, the system enters a deep quantum diffusion regime, and the picture of an electron moving along a single smooth trajectory ceases to be valid. In the Bohr model, $\Delta R = 2\pi a_0 n^2$, $\Delta T = \frac{2\pi\hbar^3}{me^4}n^3$, and $\Delta R \Delta T^{-1/2} \simeq \sqrt{2\pi}\sqrt{\frac{\hbar}{m}}n^{1/2}$, indicating that the scale of the orbital motion is somewhat larger than that of the stochastic process, suggesting that the orbital model of quantum motion had meaning only marginally (and perhaps accidentally).

9.4 Quantum Stochastic Processes and the Uncertainty Relation

Equations (5.2.19) and (5.2.20) define momentum as a function of the coordinate q and energy as a function of time, respectively, and thus may appear to violate the uncertainty relation. What is the relationship between these expressions and the uncertainty principle? Furthermore, can $\vec{p}^{\,\mathrm{local}}(q,t)$ be regarded as a hidden variable? A hidden variable is a hypothetical variable introduced under the assumption that the absence of determinism in Schrödinger dynamics, augmented by Born's probability interpretation, is due to the existence of unknown (that is, hidden) variables. As will be discussed later, even if $\vec{\nabla}\theta(q,t)$ serves as the velocity field coefficient, the accompanying Wiener process brings about dynamical uncertainly in individual dynamics (recall that each trajectory constitutes a stochastic process wandering among Bohmian trajectories). Thus, even if one calls $\vec{p}^{\,\mathrm{local}}(q,t)$ a hidden variable, classical determinism is not recovered.

Up to this point, we have seen that, due to the constraint imposed by the scaling law Eq. (9.3.9), which naturally emerges from quantum stochastic paths and path integrals, it is not possible to take the limits $\Delta q \to 0$ and $\Delta q/\Delta t \to 0$ independently. The particle is driven by the velocity field

$$\frac{1}{\rho}\begin{pmatrix}\phi_r & \phi_c\end{pmatrix}\hat{p}\begin{pmatrix}\phi_r \\ \phi_c\end{pmatrix}, \tag{9.4.1}$$

yet an intrinsic uncertainty of this kind always exists around it. Based on this, we now consider the uncertainty relation in quantum theory from a dynamical viewpoint. The uncertainty relation has historically attracted great interest as a philosophical issue related to the distinctive nature of quantum theory as a scientific framework [12, 31–35]: Initially, Heisenberg established Eq. (3.3.3) from the trade-off between the disturbance in momentum inevitably introduced by measurements at the quantum level and the precision of position measurements. Motivated by Kennard's work, Robertson and Schrödinger showed, through Eqs. (3.3.7) and (3.3.8), respectively, that position and momentum cannot be simultaneously determined as expectation values of the Schrödinger wavefunction. Ozawa unified these approaches and proposed Eq. (3.3.9), thereby refining Heisenberg's uncertainty principle.

In this way, Heisenberg's uncertainty principle and the Kennard-Robertson-Schrödinger (KRS) uncertainty relations are of different origins. Unlike Heisenberg's formulation, the KRS uncertainty relations are mathematically universal statements concerning properties of distributions and operator noncommutativity, and thus, the physical mechanism of quantum uncertainty is left implicit. In what follows, we consider the uncertainty from a different perspective based on the scaling law arising from stochastic processes of quantum paths, rather than from the KRS perspective.

9.4.1 Uncertainty Relation Between Momentum and Position

The Δq appearing in the scaling law of Eq. (9.3.9) cannot be interpreted, unlike in the case of a Brownian particle, as the distance shifted during a time interval Δt as a result of kicks by the surrounding particles of a medium. By contrast, the uncertainty arises from a mathematical structure of the Schrödinger equation and the path integral, irrespective of physical origin possible. Precisely for this reason, reconstruction of the Schrödinger equation and the real-valued path integral from scratch with mathematical rigor are all the more important.

Here, based on the fact that the scaling law has been formulated from the skew Wiener measure of the real path integral, Eq. (9.3.7), we adopt the interpretation that "during a time interval Δt, the particle is, on average, confined within a region of size Δq (in each dimension)," or equivalently, "during Δt, the particle is, on average, not found outside a region of size Δq determined by Eq. (9.3.9)." Of course, this description applies when the magnitude of the velocity field is zero. When the velocity field coefficient is finite, as in Eq. (9.4.1), the quantity Δq should be defined relative to the endpoint of the motion predicted by that field. Therefore, if one wishes to capture the particle position more precisely with a camera, the shutter time (i.e., the shutter speed) Δt must be made as short as possible. For example, if one wishes to improve the positional accuracy by a factor of 10, $\Delta q \rightarrow \Delta q/10$, then the shutter speed must be increased by a factor of 100, $\Delta t \rightarrow \Delta t/100$. Conversely, making the shutter speed arbitrarily fast, i.e., taking $\Delta t \rightarrow 0$, does not allow $\Delta q \rightarrow 0$ to be achieved on average.

Next, consider a situation in which a particle is at $q_t = X_t$ at time t and is driven by the local momentum p^{local} that defines the velocity field coefficient in Eq. (8.2.26)

$$p^{\mathrm{local}} = \frac{1}{\rho} \left(\phi_r \; \phi_c \right) \hat{p} \begin{pmatrix} \phi_r \\ \phi_c \end{pmatrix}. \tag{9.4.2}$$

In what follows, we simply write $p_t = p^{\mathrm{local}}$. We then say that the particle is at (q_t, p_t). From this point, when considering the position $X_{t+\Delta t}$ after a short time interval Δt, the particle acquires an uncertainty width Δq measured from the dynamically evolved position $(q_{t+\Delta t}, p_{t+\Delta t})$. It should be noted that the magnitude of Δq does not depend on (q_t, p_t). In this motion, in addition to $p_{t+\Delta t}$, there appears to be, effectively, an additional momentum with a width of $m\Delta q/\Delta t$. We write this as

$$\Delta p = m\Delta q/\Delta t \tag{9.4.3}$$

Of course, since Δq represents a deviation from $q_{t+\Delta t}$, it can take both positive and negative values, and the same applies to Δp. Therefore, redefining $\Delta q \equiv |\Delta q|$ and $\Delta p \equiv |\Delta p|$, the scaling law can be reinterpreted as

$$\langle \Delta p \Delta q \rangle = \hbar. \tag{9.4.4}$$

This represents the average area of a rectangle with sides Δp and Δq.

In practice, the particle may move in a zigzag manner within the width Δq, covering a longer path within the same time interval and thereby acquiring a larger average momentum. Thus, the right-hand side of Eq. (9.4.4) should be regarded as a lower bound, leading to the inequality

$$\langle \Delta p \Delta q \rangle \geq \hbar. \tag{9.4.5}$$

(For zigzag paths and their fractal nature in path integrals, see the well-known works by Suzuki et al. [36–38].)

In Eq. (9.4.4), one might be tempted to think that decreasing Δt reduces Δq and hence also reduces the momentum, so that the product Δq and Δp becomes smaller. However, in reality, if Δq is scaled as $\Delta q \to \Delta q/N$, then the time must scale as $\Delta t \to \Delta t/N^2$, and consequently, $\Delta p \to N\Delta p$. Therefore, the product $\Delta q \times \Delta p$ remains unchanged. In this way, the properties of the stochastic process prevent us from observing detailed events within an area smaller than $\langle \Delta p \Delta q \rangle = \hbar$. This is consistent with the common sense of quantum theory.

It should be noted that, according to Eq. (5.2.16), integrating $p^{\mathrm{local}}(q, t)$ over the entire space with the weight $\rho(q, t)$ yields the expectation value of momentum. Thus, $p^{\mathrm{local}}(q, t)$ represents local information at each space-time point (q, t), independent of the distribution density $\rho(q, t)$. On the other hand, in the stochastic differential equation for the path X_t, the velocity field coefficient $\alpha^{\mathrm{Real}}(X_t, t) = p^{\mathrm{local}}(q, t)/m$ is normalized by $\rho(q, t)$ (i.e., divided by it), and the resulting $\Delta p \Delta q$ represents the intrinsic uncertainty (fluctuation) associated with each individual quantum path, regardless of where it is located. Therefore, averaging these quantities with the weight $\rho(q, t)$ according to Eq. (5.2.16) yields Eq. (9.4.4). In this sense, the uncertainty relation (9.4.5) originates from quantum stochastic path dynamics.

Now, in the Kennard-Robertson-Schrödinger (KRS) uncertainty relation, momentum is from the outset defined as the expectation value $\langle \hat{p} \rangle$ in Eq. (5.2.16). Let us therefore compare Eq. (9.4.5) with Kennard's uncertainty relation:

$$\sigma(Q)\sigma(P) \geq \frac{\hbar}{2} \tag{9.4.6}$$

Here, $\sigma(Q)$ denotes the uncertainty in position as the standard deviation of q, and $\sigma(P)$ denotes the uncertainty in momentum as the standard deviation of p. Both quantities are directly obtained from the Schrödinger wavefunction. Notably, the right-hand side of Eq. (9.4.6) is $\hbar/2$, which is half of $\hbar$ in Eq. (9.4.5). This difference has a meaningful origin.

To capture the meaning of the uncertainty relation (9.4.6) further, let us briefly review the well-known method often referred to as Heisenberg's second approach. To examine the spatial spread of the Schrödinger wavefunction and its standard deviation, consider the Gaussian function

$$\psi(q, t) = N \exp\left(-\frac{\alpha}{2\hbar}(q - q_0)^2 + \frac{i}{\hbar} p_0 (q - q_0)\right). \tag{9.4.7}$$

Its probability density function is

$$|\psi(q, t)|^2 = const. \exp\left(-\frac{\alpha}{\hbar}(q - q_0)^2\right), \tag{9.4.8}$$

from which the standard deviation of the spatial spread is obtained as

$$\sqrt{(\Delta q)^2} = \sqrt{\frac{\hbar}{2\alpha}}. \tag{9.4.9}$$

Similarly, its momentum representation is

$$\begin{aligned} \tilde{\psi}(p, t) &= \int dq \, \langle p \, | q \rangle \, \langle q \, | \psi \rangle \\ &= N \exp\left(-\frac{i}{\hbar} p_0 q_0\right) \exp\left[-\frac{1}{2\hbar\alpha}(p - p_0)^2\right] \end{aligned} \tag{9.4.10}$$

and the corresponding probability density in momentum space is

$$\left|\tilde{\psi}(p, t)\right|^2 = const. \exp\left(-\frac{1}{\hbar\alpha}(p - p_0)^2\right) \tag{9.4.11}$$

with standard deviation

$$\sqrt{(\Delta p)^2} = \sqrt{\frac{\hbar\alpha}{2}}. \tag{9.4.12}$$

Thus, taking the product of Eqs. (9.4.9) and (9.4.12), the parameter α cancels, yielding the universal relation:

$$\sqrt{(\Delta q)^2}\sqrt{(\Delta p)^2} = \frac{\hbar}{2} \tag{9.4.13}$$

This was generalized to the inequality (9.4.6) by Kennard, Robertson, and Schrödinger (see Eq. (3.3.7)).

The above formal derivation of Eq. (9.4.13) clearly suggests that the subtle difference between Eq. (9.4.13) and Eq. (9.4.4) reflects a difference in how uncertainty is measured. In Kennard's uncertainty relation, $\sigma(Q)$ and $\sigma(P)$ are defined independently and can be measured separately. That is, one may measure only the uncertainty in position, tabulate the results, and then compute $\sigma(Q)$ statistically; the same applies to momentum. The product of these independently obtained quantities is then bounded by inequality (9.4.6). In contrast, the inequality (9.4.5) is based on the fact that Δq and Δp cannot be treated independently from the outset. Therefore, to measure the uncertainty in Eq. (9.4.5), one requires simultaneous measurement methods that preserve correlations, along with appropriate data processing. In such cases, the uncertainty widths directly associated with correlated Δq and Δp are found to be larger than in the KRS case.

Uncertainty relations arising from noncommuting operators are mathematically characteristic of quantum theory. However, the uncertainty relation in Eq. (9.4.13) reflects a universal property between a unimodal function with spatial spread and its Fourier transform, and it is difficult to extract the dynamical origin (mechanism) of quantum uncertainty solely from such relations, including the KRS formulation. (Uncertainty relations also appear in various other fields, such as wave theory and statistical mechanics, and are not exclusive to quantum mechanics.) On the other hand, Heisenberg's uncertainty principle, Eq. (3.3.3), formulates the physical principle that the act of observation necessarily introduces a certain disturbance into the physical process of the system being observed. Although the uncertainty relation in Eq. (9.4.5) does not describe measurement-induced disturbance, it emerges inevitably from the physical background of quantum stochastic processes. Therefore, if this uncertainty can be measured independently of other types of uncertainty and thereby Eq. (9.4.5) can be verified on its own, it would provide experimental confirmation that the stochastic process described by Eq. (8.2.26), which is considered to capture the essence of quantum dynamics, represents a physical reality.

Since the uncertainty relation follows from the scaling law Eq. (9.3.11), another uncertainty relation should also arise from the scaling law of Brownian motion itself, Eq. (9.3.4). Indeed, starting from Eq. (9.4.7) with $p_0 = 0$, one can easily arrive at the uncertainty relation along the same line of reasoning. However, it is algebraically impossible to derive uncertainty relations involving noncommuting operators, such as Eq. (3.3.7), within this framework. In this context, Koide and Kodama, in their paper titled "Generalization of uncertainty relation for quantum and stochastic systems" [39], start from classical averages and variances of position and momentum and, using the stochastic variational method developed by Yasue [40], derive Kennard's standard deviation. This work is considered to provide an important clue toward the physical foundation of the KRS uncertainty relation.

9.4.2 Uncertainty Relation Between Time and Kinetic Energy

Next, we turn to the uncertainty relation between energy and time:

$$\Delta E \times \Delta t \sim \hbar. \tag{9.4.14}$$

Various interpretations and theoretical formulations have been proposed for the energy-time uncertainty relation. The famous Einstein-Bohr debate is too specialized to discuss here, so we set it aside [35]. In general, the interpretation concerns the quantum-mechanical property relating "the precision of energy transferred between two systems, or between two orthogonal states, under perturbation, and the duration of the measurement." The textbook by Landau and Lifshitz states that "the conservation of energy can only be verified within an accuracy of $\hbar/\Delta t$ by two measurements" [31]. What is important here is that Δt is a parameter that can be chosen by the observer and does not represent fluctuations or disturbances in the measured value; thus, it is essentially different from the uncertainty relation between position and momentum. Furthermore, Mandelstam and others developed the interpretation that the uncertainty in ΔE and Δt is related to the "rate" of change of energy in a system [41], while Margolus and Levitin argued that it concerns the upper bound on the rate of energy exchange between systems [42]. There are also theories that geometrize the distance between two systems exchanging energy and measure geodesics [43]. Under the fundamental question of what a "clock" is that can measure time microscopically, Salecker and Wigner argued that "the minimum uncertainty in the mass of a clock is given by Heisenberg's relation (9.4.14)" [44]. For a more detailed discussion, see Shogo Tanimura's review article "Diversifying Uncertainty Relations" (2016) (available in the Nagoya University Academic Repository: https://nagoya.repo.nii.ac.jp).

In this book, we do not (or cannot) enter into these deep debates. Instead, we transform the scaling law, Eq. (9.3.9), into an uncertainty relation between kinetic energy and time. This is entirely different from the energy-time uncertainty relation discussed above, which concerns "the precision of energy transfer and its rate." Fundamentally, there is no Schrödinger wavefunction that represents a time width, and time, as a parameter, does not possess a self-adjoint operator. (For the same reason, careful discussion is also required for the uncertainty relation between angular momentum and rotation angle [45, 46].) Therefore, time does not admit a standard deviation such as $\sigma(Q)$ or $\sigma(P)$, and it cannot be expressed in a form like Eq. (9.4.6).

We resume from the scaling law Eq. (9.3.9). First, we rewrite it equivalently as

$$\left\langle \frac{m}{2} (\frac{\Delta q}{\Delta t})^2 \Delta t \right\rangle = \frac{\hbar}{2}. \tag{9.4.15}$$

Here, $m(\Delta q/\Delta t)^2/2 = \Delta E_K$ may be regarded as the fluctuation of kinetic energy that appears during the interval Δt. Thus, Eq. (9.4.15) can be rewritten as

$$\langle \Delta E_K \Delta t \rangle = \frac{\hbar}{2}. \tag{9.4.16}$$

Again, since the particle may undergo zigzag motion within the given widths Δq and Δt, the kinetic energy may take larger values, and we replace this with the inequality:

$$\langle \Delta E_K \Delta t \rangle \geq \frac{\hbar}{2} \tag{9.4.17}$$

Here, as well, under the scaling $\Delta t \rightarrow \Delta t/N$, we have $\Delta q \rightarrow \Delta q/\sqrt{N}$ and $\Delta E_K \rightarrow N \Delta E_K$, showing that Eq. (9.4.16) is invariant under such rescaling. Incidentally, the kinetic energy in Eq. (9.4.17) is simply defined, for $\Delta t = t_2 - t_1$, as

$$\Delta E_K = \frac{m}{2} \left(\frac{q(t_2) - q(t_1)}{t_2 - t_1} \right)^2. \tag{9.4.18}$$

Again, we stress that the uncertainty relation between kinetic energy and time obtained here is entirely different from the historically deep discussions of the energy-time uncertainty relation mentioned above and is rather simple. It is merely an equivalent transformation of the uncertainty relation between momentum and position discussed in the previous section, and its physical background is clear in that it is derived from the scaling law of quantum stochastic processes. One may say that "a long time is required to measure kinetic energy accurately," or, if kinetic energy is regarded as a "local temperature," it may also be interpreted as an uncertainty relation between temperature measurement and time.

9.4.3 Meaning of Uncertainty in Stochastic Processes

The uncertainty relations formulated here have a completely different origin from Heisenberg's uncertainty principle and the KRS uncertainty relations, and it is worthwhile to reconsider their meaning.

In connection with the KRS inequality, the commonly shared interpretations are as follows:

(1) Noncommuting observables A and B cannot simultaneously possess definite values. (Moreover, Kochen and Specker established a stronger theorem: "it is impossible to assign definite values to all physical observables simultaneously" [47].)
(2) The preparation for measurements of these A and B can only be carried out exclusively (if one reduces the standard deviation of one, the other increases), and the result depends on the measurement method.
(3) It is fundamentally impossible to know what values the system possesses prior to measurement; therefore, one can only speak about quantities that can actually be measured.

These points indicate a profound gap between quantum theory and our classical way of recognition.

On the other hand, the author interprets the uncertainty relations derived from quantum stochastic processes in this chapter as follows:

(1) A restriction on the concept of infinitesimal in physics. For example, $\Delta q \to 0$ and $\Delta p \to 0$ are not independent. This is reminiscent of Lorentz transformations, although no direct connection is known at present.
(2) Not only are there constraints between them, but there also exists an intrinsic (averaged) functional relation for position and momentum

$$F_1(\Delta q, \Delta p) = 0, \tag{9.4.19}$$

and another form of functional relation

$$F_2(\Delta E_K, \Delta t) = 0 \tag{9.4.20}$$

for kinetic energy and time.
(3) Consequently

$$\lim_{\Delta t \to 0,\, \Delta q \to 0} \frac{\Delta q}{\Delta t} \tag{9.4.21}$$

and $\Delta q \to 0$ cannot coexist, and even if one constructs an implicit function $q(t)$, it loses smoothness almost everywhere. Similarly, if one considers the kinetic energy as an implicit function of time, $\Delta E_K(t)$, it is continuous but not smooth at almost all times.
(4) Equation (9.4.19) conflicts with the definition of "physical reality" deeply discussed in the EPR argument [48], which we quote here again "If, without in any way disturbing a system, we can predict with certainty (i.e., with probability equal to unity) the value of a physical quantity, then there exists an element of physical reality corresponding to this physical quantity." However, these intrinsic relations like F_1 and F_2 exist independently of observation. That is,

the physical reality may lie in the functional relations such as F_1 and F_2 rather than the individual variables.

(5) The uncertainty relations arising from this stochastic process are intrinsic kinematical properties that exist independently of observation. (They are unrelated to Heisenberg's uncertainty principle.) However, in the measurement processes of quantities such as Δq and Δp, the experimental procedures used to obtain the outputs (observed values) may themselves depend on the functions F_1 and/or F_2. In other words, even if the functional relation F_1 (F_2) itself is unique, the way its effects appear in the output may vary depending on the measurement conditions.

9.5 Quantum Entanglement and Spontaneous Disentanglement on Quantum Paths

As discussed in Chap. 3, Sect. 3.3.5, quantum entanglement is, for researchers who apply the Schrödinger equation to many-body problems, a consequence of fundamental requirements and is not regarded as particularly problematic. For example, in order to describe the electronic states of chemical bonds in molecules using a Schrödinger function, it constitutes one of the necessary underlying "premises."

However, in light of the fact that entanglement was taken by Einstein and others as evidence that quantum theory is either incorrect or incomplete and that it has led to prolonged debates, we shall here consider somewhat deeper aspects of entanglement from the standpoint of quantum stochastic path dynamics.

In general, entanglement is discussed in the context of the superposition principle of the Schrödinger function. For example, a two-particle Schrödinger function is said to be entangled when it can only be expressed as a sum of products of single-particle functions:

$$\psi(q_1, q_2, t) = \sum_i c_i(t)\chi_i(q_1)\xi_i(q_2). \tag{9.5.1}$$

However, as time evolves, the state may converge to a form such as

$$\psi(q_1, q_2, t) \to c_1(t)\chi_1(q_1)\xi_2(q_2), \tag{9.5.2}$$

in which case it does not present a difficulty of the kind that troubled Einstein. The problem arises when an underlying symmetry, for example, prevents resolution into a single product form, even when the distance between q_1 and q_2 becomes arbitrarily large. The implications of such non-separability have already been briefly summarized in Chap. 3, Sect. 3.3.5.

As has been emphasized repeatedly, as long as one remains confined to the Schrödinger equation alone, the peculiar nature of entanglement cannot be fully resolved. Therefore, in what follows, we shall consider, through specific examples, what may occur within the framework of quantum stochastic process dynamics.

9.5.1 Entanglement in the Velocity Field Coefficient

Let us first examine how entanglement appears in the quantum stochastic differential equation (8.2.26). For a two-electron system, considering a single stochastic path is sufficient for the present purpose. As an example of a Schrödinger function, we take

$$\psi(Q_1, Q_2, t) \\ = N(t)\,[a(q_1)b(q_2) + b(q_1)a(q_2)]\,\frac{1}{\sqrt{2}}\,[\alpha(\omega_1)\beta(\omega_2) - \beta(\omega_1)\alpha(\omega_2)]\,. \tag{9.5.3}$$

The coordinates of a single electron are described by four variables, $Q_1 = (x_1, y_1, z_1, \omega_1)$. Here, $q = (x, y, z)$ denotes the spatial coordinates, and ω represents the spin coordinate. $a(q)$ and $b(q)$ are spatially localized function at q. The essential point is that $\psi(Q_1, Q_2, t)$ possesses the exchange antisymmetry:

$$\psi(Q_2, Q_1, t) = -\psi(Q_1, Q_2, t), \tag{9.5.4}$$

which we shall use throughout the following discussion.

We assume that the potential function $V(q_1, q_2)$ appearing in the Hamiltonian does not include spin-dependent interactions, such as those arising from Coulomb interactions, and is symmetric under the exchange of q_1 and q_2:

$$V(q_2, q_1, t) = V(q_1, q_2, t). \tag{9.5.5}$$

Accordingly, the Hamiltonian operator $\hat{H}(q_1, q_2, t)$ also satisfies

$$\hat{H}(q_2, q_1, t) = \hat{H}(q_1, q_2, t). \tag{9.5.6}$$

As already noted, the singlet spin function contained in $\psi(Q_1, Q_2, t)$, that is

$$\frac{1}{\sqrt{2}}\,[\alpha(\omega_1)\beta(\omega_2) - \beta(\omega_1)\alpha(\omega_2)]\,, \tag{9.5.7}$$

is itself entangled. It is known that Bell's inequality can be shown to be violated using only the spin degrees of freedom; however, we shall leave such proofs to other references (for example, A. Shimizu, *Foundations of Quantum Mechanics*, Science-sha). The proof is rigorous, yet our concern here is that if one focuses solely on the spin functions when discussing entanglement, one may overlook the possibility of time evolution of entanglement, regardless of how far apart q_1 and q_2 are and regardless of changes in the spatial part of $\psi(Q_1, Q_2, t)$ of Eq. (9.5.3):

$$\Phi(q_1, q_2) = a(q_1)b(q_2) + b(q_1)a(q_2). \tag{9.5.8}$$

Let us add another remark regarding the symmetry of $\psi(Q_1, Q_2, t)$ in Eq. (9.5.3). The spatial function extracted as in Eq. (9.5.8) is symmetric under exchange of q_1 and q_2, and since the Hamiltonian does not include spin interactions, the α-spin and β-spin components appear with equal probability. In other words, any phenomenon occurring for the α-spin can equally occur for the β-spin upon exchange.

In order to simplify the subsequent discussion, we adopt the polar representation of the Schrödinger vector:

$$\bar{\psi}(Q_1, Q_2, t) = \rho^{1/2}(Q_1, Q_2, t) \begin{pmatrix} \cos(\theta(Q_1, Q_2, t)) \\ \sin(\theta(Q_1, Q_2, t)) \end{pmatrix}. \tag{9.5.9}$$

In this representation, the density satisfies

$$\rho(Q_2, Q_1, t) = \rho(Q_1, Q_2, t), \tag{9.5.10}$$

while its square root must be antisymmetric in accordance with Eq. (9.5.4):

$$\rho^{1/2}(Q_2, Q_1, t) = -\rho^{1/2}(Q_1, Q_2, t), \tag{9.5.11}$$

and the angular component must be symmetric:

$$\theta(Q_2, Q_1, t) = \theta(Q_1, Q_2, t). \tag{9.5.12}$$

If the angular part were chosen to be antisymmetric, Eq. (9.5.4) would not be satisfied within Eq. (9.5.9). To ensure that Eq. (9.5.11) holds, one may, for example, take

$$\rho^{1/2}(Q_1, Q_2, t) = \left|\rho^{1/2}(Q_1, Q_2, t)\right| \frac{q_1 - q_2}{|q_1 - q_2|}. \tag{9.5.13}$$

In this way, the dynamics of θ (Eqs. (5.2.19) and (5.2.20)) remain symmetric.

Based on Eq. (9.1.7), the quantum stochastic differential equation (8.2.26) can be rewritten for $X_1(t) = q_1(t)$ and $X_2(t) = q_2(t)$ as

$$X_1(t + dt) = X_1(t) + \frac{\hbar}{m}\nabla_1\theta(Q_1, Q_2, t)dt + \sqrt{\frac{\hbar}{m}}dW_0^{(1)}(t, \omega), \tag{9.5.14}$$

and

$$X_2(t + dt) = X_2(t) + \frac{\hbar}{m}\nabla_2\theta(Q_1, Q_2, t)dt + \sqrt{\frac{\hbar}{m}}dW_0^{(2)}(t, \omega). \tag{9.5.15}$$

Since we are dealing with electrons, we set $m_1 = m_2 = m$. Here, we have already used the symmetry property of $\theta(X_1, X_2, t)$ of Eq. (9.5.12). In this dynamics, only the angular variable $\theta(X_1, X_2, t)$ appears, while ρ does not explicitly appear. The reason for the absence of ρ has been discussed several times (see, for example, Eq. (8.2.33)).

In the coupled equations of (9.5.14) and (9.5.15), $\theta(Q_1, Q_2, t)$ serves to maintain the permutation symmetry between $X_1(t)$ and $X_2(t)$, whereas the two independent Wiener processes $W_0^{(1)}(t, \omega)$ and $W_0^{(2)}(t, \omega)$ tend to break it. However, the effect of symmetry breaking must be small as long as the magnitude of the velocity field coefficients is large enough.

To examine the content of $\theta(X_1, X_2, t)$, we return to Eq. (9.1.10), which gives

$$\frac{\hbar}{m}\nabla_i\theta(Q_1, Q_2, t) = \frac{1}{\rho m}P_{X_t}^{(i)}(Q_1, Q_2) \quad (i = 1, 2), \tag{9.5.16}$$

and the time derivative of $P_{X_t}^{(i)}(X_1, X_2)$ follows Eq. (9.1.9) such that

$$\begin{aligned} \frac{dP_{X_t}^{(i)}}{dt} &= -\rho(Q_1, Q_2, t)\nabla_i V(q_1, q_2) \\ &= -\nabla_i V(q_1, q_2) \times |N(t)|^2 \\ &\Big[|a(q_1)|^2\,|b(q_2)|^2 + |b(q_1)|^2\,|a(q_2)|^2 \\ &- \big(a^*(q_1)b(q_1)b^*(q_2)a(q_2) + \text{c.c.}\big)\Big] \\ &\times \frac{1}{2}\left[\alpha(\omega_1)\beta(\omega_2) - \beta(\omega_1)\alpha(\omega_2)\right]^2, \end{aligned} \tag{9.5.17}$$

with "c.c." denoting the complex conjugate. Since the spin part does not affect the dynamics, it can be integrated out, leading effectively to

$$\begin{aligned} \frac{dP_{X_t}^{(i)}}{dt} &= -\rho(Q_1, Q_2, t)\nabla_i V(q_1, q_2) \\ &= -\nabla_i V(q_1, q_2) \times |N(t)|^2 \\ &\Big[|a(q_1)|^2\,|b(q_2)|^2 + |b(q_1)|^2\,|a(q_2)|^2 \\ &- \big(a^*(q_1)b(q_1)b^*(q_2)a(q_2) + \text{c.c.}\big)\Big]. \end{aligned} \tag{9.5.18}$$

Because the spatial function of $\psi(Q_1, Q_2, t)$ has been chosen as in Eq. (9.5.8), the trajectories $X_1(t) = q_1(t)$ and $X_2(t) = q_2(t)$ are completely unaffected by spin. In this expression, the term

$$a^*(q_1)b(q_1)b^*(q_2)a(q_2) + \text{c.c.} \tag{9.5.19}$$

is the so-called interference term, which reveals the effect of entanglement and its physical meaning. Equation (9.5.18) is, of course, symmetric under the exchange of q_1 and q_2.

In this way, since the velocity field is symmetric under particle exchange, integrating Eqs. (9.5.14) and (9.5.15) simultaneously preserves the exchange symmetry between Q_1 and Q_2. On the other hand, within the Schrödinger equation, this exchange symmetry is strictly conserved. This is because the Hamiltonian (Eq. (9.5.6)) commutes with the particle exchange operator, and thus, there exists no mathematical mechanism that would violate the property in Eq. (9.5.4). For complement, we note that

$$i\hbar\frac{\partial}{\partial t}\psi(Q_1, Q_2, t) = \left(-\frac{\hbar^2}{2m}(\nabla_1^2 + \nabla_2^2) + V(q_1, q_2)\right)\psi(Q_1, Q_2, t) \quad (9.5.20)$$

is equivalent to

$$i\hbar\frac{\partial}{\partial t}(-\psi(Q_2, Q_1, t)) = \left(-\frac{\hbar^2}{2m}(\nabla_1^2 + \nabla_2^2) + V(q_1, q_2)\right)(-\psi(Q_2, Q_1, t)), \quad (9.5.21)$$

and therefore the relation between $\psi(Q_1, Q_2, t)$ and $-\psi(Q_2, Q_1, t)$ is never violated.

9.5.2 Spontaneous Disentanglement (Detanglement)

Entanglement is a state based on the principle of superposition, which is of fundamental importance in making quantum mechanics what it is. However, as aforementioned, the peculiarity perceived by Einstein lies in the fact that the entangled state does not disappear no matter how far the particles are separated. Let us now examine this point.

Consider, in the above two-electron system, the situation in which, after a sufficiently long time (formally $t \to \infty$), the two particles originally located at q_1 and q_2 become widely separated and are divided into two spatial regions, A and B. Let us denote the coordinates in regions A and B by $q(\mathrm{A})$ and $q(\mathrm{B})$, respectively, and assume that $|q(\mathrm{A}) - q(\mathrm{B})| \sim \infty$. In this case, since the Schrödinger function has no mechanism to break the exchange symmetry in Eq. (9.5.3), it is still expressed as a sum of products:

$$\begin{aligned}
&\psi^{\mathrm{asymptotic}}(Q_1, Q_2, t) \\
&= N(t)\left[a^A(q_1(\mathrm{A}))b^B(q_2(\mathrm{B})) + b^B(q_1(\mathrm{B}))a^A(q_2(\mathrm{A}))\right] \\
&\times [\alpha(\omega_1)\beta(\omega_2) - \beta(\omega_1)\alpha(\omega_2)].
\end{aligned} \quad (9.5.22)$$

That is, the state remains entangled. Here, $a^A(q)$ and $b^B(q)$ are functions localized in regions A and B, respectively ($a^A(q(\mathrm{B})) \simeq 0$, $b^B(q(\mathrm{A})) \simeq 0$). Since they are infinitely separated, one has $a^A(q(\mathrm{A}))b^B(q(\mathrm{B})) \simeq 0$, and the interference terms vanish even mathematically. Nevertheless, in Eq. (9.5.22), each electron, whether associated with q_1 or q_2, has an equal probability of being in region A or B. Moreover, the form of Eq. (9.5.22) suggests that the states $a^A(q_1(\mathrm{A}))b^B(q_2(\mathrm{B}))$ and $b^B(q_1(\mathrm{B}))a^A(q_2(\mathrm{A}))$ remain correlated.

Now, suppose that the electron observed in region A happens to have α-spin. Then, the electron in region B must necessarily have β-spin, which is consistent. However, it becomes puzzling to ask how the information that either one of the components, i.e. $a^A(q_1(\mathrm{A}))\alpha(\omega_1)b^B(q_2(\mathrm{B}))\beta(\omega_2)$ or $a^A(q_2(\mathrm{A}))\alpha(\omega_2)b^B(q_1(\mathrm{B}))\beta(\omega_1)$, has been physically selected out of Eq. (9.5.22) is transmitted to region B. An observer in region B cannot know this, since the probabilities of finding α-spin and β-spin remain equal, and no communication from region A is available (at least not faster than light). Who, then, has selected these components from the entangled Schrödinger function? (Some would describe this by saying that the observer in region A has "selected" even the component localized in the distant region B, or that the wave packet has "collapsed" accordingly.) Who provides such nonlocal information to the observer in region B? For those who think deeply, like Einstein, it is natural to conclude that quantum mechanics, which permits such nonlocal correlations, is either incomplete or lacking something essential. Although experiments confirming the violation of Bell's inequality have established the firm existence of entanglement in quantum mechanics, it may still be argued that Einstein's sincere question has not yet been fully answered.

On the other hand, as discussed in Chap. 5, the Schrödinger function is a distribution amplitude function, and in order to construct a function that represents a distribution, it is necessary at some level to take a sum over all possible events. This was the main theme of Chap. 7. From this standpoint, it is rather natural to write the distribution amplitude function in the form of Eq. (9.5.22), even in situations where $|q(\mathrm{A}) - q(\mathrm{B})| \sim \infty$. Yet, even if it is understood that the Schrödinger function becomes a distribution amplitude through the summation over a large (in fact, infinite) number of trajectories, it is difficult to accept that more than one physical event is superposed on a single trajectory described by the coupled quantum stochastic equations (9.5.14) and (9.5.15). Therefore, it is necessary to examine, within the framework of individual quantum path dynamics, whether nonlocal correlations extending to infinity truly exist as a consequence of the superposition principle.

Let us now assume that the two electrons have actually separated. Then, in this limit, the spatial part of the Schrödinger function $\Phi(q_1, q_2)$ in Eq. (9.5.8) should read in the Bohm representation as

$$\Phi(q_1, q_2) = \rho^{1/2}(q_1, q_2) \exp\left(\frac{i}{\hbar}\left(S_1(q_1) + S_2(q_2)\right)\right), \tag{9.5.23}$$

in which the part responsible for spatial distribution and that for dynamics is well separated. The dynamical part represents the quantum phase, relevant to momentum through $p_1 = \nabla S_1$ and $p_2 = \nabla S_2$, no matter which of $a(q)$ or $b(q)$ functions the electrons belong to, since they are independent now. Therefore, by comparing Eqs. (5.2.18) and (9.5.23), we see that the function $\theta(X_1, X_2, t)$ in Eqs. (9.5.14) and (9.5.15), respectively, can be expressed as a sum of independent contributions:

$$\theta(X_1(t), X_2(t), t) \to \theta_1(X_1(t), t) + \theta_2(X_2(t), t). \tag{9.5.24}$$

(Recall Eqs. (5.2.19) and (5.2.20) for the quantum dynamics through the angular part.) To make the argument more explicit, we rewrite Eqs. (9.5.14) and (9.5.15) as

$$X_1(t + dt) \to X_1(t) + \frac{\hbar}{m}\nabla_1\theta_1(X_1(t), t)dt + \sqrt{\frac{\hbar}{m}}dW_0^{(1)}(t, \omega) \tag{9.5.25}$$

and

$$X_2(t + dt) \to X_2(t) + \frac{\hbar}{m}\nabla_2\theta_2(X_2(t), t)dt + \sqrt{\frac{\hbar}{m}}dW_0^{(2)}(t, \omega). \tag{9.5.26}$$

Thus, in the asymptotic region, $X_1(t)$ and $X_2(t)$ become dynamically independent. In other words, although entanglement is preserved in the Schrödinger function as a distribution amplitude, it disappears from the dynamics of individual quantum trajectories. Therefore, disentanglement must occur somewhere in the transition from Eqs. (9.5.14) and (9.5.15) to Eqs. (9.5.25) and (9.5.26). Let us now investigate this process.

It is natural to expect that disentanglement occurs in the component of the Schrödinger function that forms the velocity field coefficient:

$$\phi_1 = a(q_1)b(q_2) + b(q_1)a(q_2), \tag{9.5.27}$$

when the interference term between the two components becomes small, and the states

$$\phi_2 = a(q_1)b(q_2), \tag{9.5.28}$$

and

$$\phi_3 = b(q_1)a(q_2) \tag{9.5.29}$$

have nearly the same energy (quasi-degenerate) and acquire similar magnitudes and directions of the local momentum $\vec{p}^{\text{local}}(q, t)$. Even if these three states are quasi-degenerate, symmetry conservation at the level of the Schrödinger equation prevents ϕ_1 from directly mixing with and transitioning to ϕ_2 or ϕ_3. On the other hand, along quantum stochastic paths, the randomess due to the stochastic processes

$\sqrt{\frac{\hbar}{m}}dW_0^{(1)}(t,\omega)$ and $\sqrt{\frac{\hbar}{m}}dW_0^{(2)}(t,\omega)$ can cause $X_1(t)$ and $X_2(t)$ to undergo small random displacements. As a result, the velocity field coefficient derived from ϕ_1 begins to be influenced by those derived from ϕ_2, ϕ_3, and their arbitrary linear combinations. Consequently, the dynamics may gradually become dominated by the velocity fields of ϕ_2 or ϕ_3, and once separation begins, the system may enter the independent mode described by Eqs. (9.5.25) and (9.5.26). (Of course, if the interaction is sufficiently strong, the system may return to the original entangled dynamics of Eqs. (9.5.14) and (9.5.15).)

With this in mind, let us reconsider Eqs. (9.5.14) and (9.5.15). In fact, these equations do not distinguish between the cases where X_1 approaches $q(\mathrm{A})$ and X_2 approaches $q(\mathrm{B})$, and where X_1 approaches $q(\mathrm{B})$ and X_2 approaches $q(\mathrm{A})$. Let us therefore denote the former case by X_1^A and X_2^B. Then, Eqs. (9.5.14) and (9.5.15) correspond to the two possible systems:

$$\begin{cases} X_1^A(t+dt) = X_1^A(t) + \frac{\hbar}{m}\nabla_1\theta_1(X_1^A(t),t)dt + \sqrt{\frac{\hbar}{m}}dW_0^{(1A)}(t,\omega) \\ X_2^B(t+dt) = X_2^B(t) + \frac{\hbar}{m}\nabla_2\theta_2(X_2^B(t),t)dt + \sqrt{\frac{\hbar}{m}}dW_0^{(2B)}(t,\omega) \end{cases} \tag{9.5.30}$$

and

$$\begin{cases} X_1^B(t+dt) = X_1^B(t) + \frac{\hbar}{m}\nabla_1\theta_1(X_1^B(t),t)dt + \sqrt{\frac{\hbar}{m}}dW_0^{(1B)}(t,\omega) \\ X_2^A(t+dt) = X_2^A(t) + \frac{\hbar}{m}\nabla_2\theta_2(X_2^A(t),t)dt + \sqrt{\frac{\hbar}{m}}dW_0^{(2A)}(t,\omega) \end{cases} \tag{9.5.31}$$

of which one case is realized in practice. Since the Wiener processes in Eq. (9.5.30) and Eq. (9.5.31) are independent, the symmetry between the path combinations $(X_1^A(t), X_2^B(t))$ and $(X_1^B(t), X_2^A(t))$ is broken. This symmetry breaking occurs spontaneously.

Let us recall here the uncertainty relations in Eqs. (9.4.5) and (9.4.17). The uncertainties relevant to momentum and kinetic energy introduce fluctuations in the velocity and kinetic energy in the quantum stochastic differential equations, thereby assisting further mixing in the velocity field as described above. In this way, quantum trajectories can undergo spontaneous disentanglement (detanglement) through intrinsic stochastic processes, rather than through external fields.

Finally, it is evident that which region $X_1(t)$ and $X_2(t)$ reach—i.e., which of the possible channels

$$(X_1(\mathrm{A})\alpha, X_2(\mathrm{B})\beta) \quad \text{or} \quad (X_2(\mathrm{A})\beta, X_1(\mathrm{B})\alpha) \tag{9.5.32}$$

or

$$(X_1(\mathrm{A})\beta, X_2(\mathrm{B})\alpha) \quad \text{or} \quad (X_2(\mathrm{A})\alpha, X_1(\mathrm{B})\beta) \tag{9.5.33}$$

is realized depends on which of Eqs. (9.5.30) or (9.5.31) is randomly selected. The channel is determined by the stochastic nature of the Wiener processes in these equations.

The above analysis suggests that if entanglement has an "exit," then a corresponding mechanism must also exist for its "entrance." It is therefore natural to consider that two or more particles that are initially unentangled may approach each other and enter an entangled state. However, this aspect has received little attention so far.

9.5.3 Disentanglement and Bell's Inequality

It is often said that the experimental verification of the violation of Bell's inequality constitutes one of the greatest achievements in elucidating the essential strangeness of quantum mechanics. At times, it is even claimed that "Einstein was wrong." In reality, however, Einstein regarded the peculiar nature of entanglement as a challenge to the locality and separability of physical laws. Therefore, even the conclusion that "Bell's inequality is violated," namely, that "local hidden variables do not exist," cannot be regarded as a complete answer to Einstein's concerns. Since it is certain that Einstein acknowledged the existence of entangled states both mathematically and physically, the author believes that his concerns can now be addressed, precisely because the possibility of detanglement within quantum stochastic path dynamics has become clearer. Conversely, even if one were to further pursue Bell's inequality in depth, it is not itself a theory of quantum mechanics and therefore cannot reach the deeper truths of quantum theory beyond a certain point. (It should be noted, however, that the author does not claim competence in discussing entanglement in cosmology; the present discussion is restricted to nonrelativistic quantum mechanics.) Nevertheless, it is necessary to examine whether the experimental violation of Bell's inequality is compatible with the notion of detanglement.

First, let us reconfirm that the final channel selection resulting from detanglement—namely, in the above example, which of the four possible combinations, Eqs. (9.5.32) and (9.5.33), is realized—is determined statistically by completely spontaneous and uncontrollable randomness arising from the Wiener processes in quantum stochastic path dynamics. On the other hand, if disentanglement does not occur and the outcome is determined solely by the Schrödinger equation and the Schrödinger function, then the selection among the channels predicted by

$$\left|a^A(q_1(\mathrm{A}))\right|^2 |\alpha(\omega_1)|^2 \left|b^B(q_2(\mathrm{B}))\right|^2 |\beta(\omega_2)|^2$$

or

$$\left|a^A(q_1(\mathrm{A}))\right|^2 |\beta(\omega_1)|^2 \left|b^B(q_2(\mathrm{B}))\right|^2 |\alpha(\omega_2)|^2$$

or

$$\left|b^B(q_1(\mathrm{B}))\right|^2 |\alpha(\omega_1)|^2 \left|a^A(q_2(\mathrm{A}))\right|^2 |\beta(\omega_2)|^2$$

or

$$\left|b^B(q_1(\mathrm{B}))\right|^2 |\beta(\omega_1)|^2 \left|a^A(q_2(\mathrm{A}))\right|^2 |\alpha(\omega_2)|^2$$

is entirely statistical. That is, if one solves the stochastic differential equations and then build the statistical distribution of the solutions, the result is indistinguishable from the probability distribution determined by the Schrödinger equation.

Thus, in observations performed in the asymptotic region, unless an additional measurement process is introduced that can directly verify whether disentanglement has occurred, the two types of channel selection are completely indistinguishable in practice. In other words, even in the presence of disentanglement, that is, the spontaneous breakdown of the once entangled state, Bell's inequality is violated anyway in exactly the same statistical manner. Since Bell's inequality itself does not refer to the internal structure of quantum theory, its role effectively ends at this point.

Furthermore, regarding the interpretation that measurement itself breaks entanglement and thereby selects one of the four possible outcomes above or induces wave function collapse, there is no need to invoke such notions. As emphasized repeatedly, the Schrödinger function is a distribution amplitude function; it is not a physical entity but rather a mathematical one, and therefore, it does not "collapse."

In this way, it becomes clear once again that the paradoxes related to entanglement posed by Einstein and others [48] arise from considering quantum states solely on the basis of the Schrödinger equation and the Schrödinger function. Conversely, it does not make sense to construct a counterexample to the uncertainty relation by exploiting long-range correlations due to entanglement. What we have understood here is that while the spontaneous selection of channels, as in Eq. (3.3.18), cannot be described without quantum stochastic paths, processes that seemingly appear unphysical, such as Eq. (3.3.19), are nevertheless valid as a solution of the Schrödinger equation along with the probabilistic interpretation.

Of course, this discussion does not imply that the full nature of entanglement has been completely clarified. Here, we have merely argued that entanglement must possess both an "entrance" and an "exit." It is beyond doubt that further development of research based on quantum stochastic path dynamics is required.

9.6 Summary of This Chapter

Since the Schrödinger function is a distribution amplitude function representing a coherent superposition of multiple events, a path dynamics theory is required when one attempts to extract individual events from it. Individual motions must, of course, be consistent with the overall description. However, the distribution function may miss (mathematical) features that can disappear when summing over individual events. Therefore, it should be clearly recognized that the wavefunction does not describe the entirety of the quantum picture. For example,

(1) The dynamics of individual paths involve stochastic processes, much like Brownian motion, but these are not due to external forces; rather, they are intrinsic to quantum dynamics. Since the Wiener process disappears upon summation over paths, it does not explicitly appear in either the diffusion equation or the Schrödinger equation.
(2) Individual paths do not possess time-reversal symmetry due to the presence of the Wiener process. However, upon averaging (taking the sum), this asymmetry vanishes, and the Schrödinger equation exhibits time-reversal symmetry as an idealized limit.
(3) Individual paths bear uncertainty relations in a kinematical sense due to the Wiener process. These are mathematically distinct from the uncertainty relations arising from the commutation relations of operators acting on the Schrödinger wavefunction (the KRS uncertainty relations), but physically they could be one of their fundamental origins.
(4) The Schrödinger equation has the important property that its solutions include entangled states; however, it does not contain a mechanism that describes when/how uncorrelated particles become entangled or when/how they disentangle. In this book, based on the theory of quantum stochastic paths, the concept of spontaneous disentanglement (detanglement) and its mathematical mechanism have been proposed. Nevertheless, proving the existence of a more universal form of detanglement and clarifying its limitations remain important tasks for future research, particularly in connection with quantum information theory.

References

1. Sanz, Á. S. & Miret-Artés, S. *A Trajectory Description of Quantum Processes. I and II. Applications: A Bohmian Perspective* (Springer, 2013).
2. Bohm, D. A suggested interpretation of the quantum theory in terms of " hidden" variables. I. *Physical Review* **85**, 166 (1952).
3. De Gosson, M. A. *The Principles Of Newtonian And Quantum Mechanics, The: The Need For Planck's Constant, h* (World Scientific, 2016).
4. Philippidis, C., Dewdney, C. & Hiley, B. J. Quantum interference and the quantum potential. *Nuovo Cimento B* **52**, 15–28 (1979).
5. Wyatt, R. E. *Quantum Dynamics with Trajectories: Introduction to Quantum Hydrodynamics*, vol. 28 (Springer Science & Business Media, 2005).
6. Miller, W. H. Classical-limit quantum mechanics and the theory of molecular collisions. *Advances in Chemical Physics* 69–177 (1974).
7. Schulman, L. S. *Techniques and Applications of Path Integration* (Courier Corporation, 2012).
8. Brack, M. & Bhaduri, R. *Semiclassical Physics* (CRC Press, 2018).
9. Takahashi, S. & Takatsuka, K. Towards many-dimensional real-time quantum theory for heavy-particle dynamics. I. Semiclassics in the lagrange picture of classical phase flow. *Physical Review A* **89**, 012108 (2014).

10. Nelson, E. Derivation of the schrödinger equation from newtonian mechanics. *Physical Review* **150**, 1079 (1966).
11. Schiff, L. I. *Quantum Mechanics* (McGraw–Hill, New York, 1968).
12. Messiah, A. *Quantum Mechanics* (Courier Corporation, 2014).
13. Wigner, E. On the quantum correction for thermodynamic equilibrium. *Physical Review* **40**, 749 (1932).
14. Zachos, C., Fairlie, D. & Curtright, T. *Quantum Mechanics in Phase Space: An Overview with Selected Papers* (World Scientific, 2005).
15. Polkovnikov, A. Phase space representation of quantum dynamics. *Annals of Physics* **325**, 1790–1852 (2010).
16. Lichtenberg, A. J. & Lieberman, M. A. *Regular and Chaotic Dynamics*, vol. 38 (Springer Science & Business Media, 2013).
17. Nakamura, K. *Quantum Chaos* (Cambridge University Press, Cambridge, UK, 1993).
18. Reichl, L. E. *The Transition to Chaos* (Springer, Berlin, 1992).
19. Gaspard, P., Alonso, D. & Burghardt, I. *New Way of Understanding Semiclassical Quantization*, 105 (John Wiley, New York, 1995).
20. Robnik, E., Marco. *Quantum Chaos. Dedicated to Professor Giulio Casati on the Occasion of His 80th Birthday* (MDPI, Basel, 2023).
21. Takatsuka, K. Quantum chaos in the dynamics of molecules. *Entropy,* **25**, 63 (2023).
22. Nagasawa, M. Transformations of diffusion and Schrödinger processes. *Probability Theory and Related Fields* **82**, 109–136 (1989).
23. Nagasawa, M. *Schrödinger Equations and Diffusion Theory* (Springer Science & Business Media, 2012).
24. Mandelbrot, B. B. *The Fractal Geometry of Nature/Revised and enlarged edition* (1983).
25. Oksendal, B. *Stochastic Differential Equations: An Introduction with Applications* (Springer Science & Business Media, 2013).
26. Einstein, A. Zum quantensatz von Sommerfeld und Epstein. *Verh. d. D. Physik. Ges* 82–92 (1917).
27. Gutzwiller, M. C. Energy spectrum according to classical mechanic. *J. Math. Phys.* **11**, 1791 (1970).
28. Gutzwiller, M. C. Periodic orbits and classical quantization conditions. *J. Math. Phys.* **12**, 343 (1971).
29. Kleinert, H. *Path Integrals in Quantum Mechanics, Statistics, Polymer Physics, and Financial markets* (World Scientific Publishing Company, 2006).
30. Heisenberg, W. *et al. Physics and beyond* (Allen & Unwin London, 1971).
31. Landau, L. D. & Lifshitz, E. M. *Quantum Mechanics: Non-Relativistic Theory*, vol. 3 (Elsevier, 2013).
32. Ozawa, M. Universally valid reformulation of the Heisenberg uncertainty principle on noise and disturbance in measurement. *Physical Review A* **67**, 042105 (2003).
33. Ozawa, M. Heisenberg's original derivation of the uncertainty principle and its universally valid reformulations. *Current Science* 2006–2016 (2015).
34. Auletta, G. *Foundations and Interpretation of Quantum Mechanics: In the Light of a Critical-Historical Analysis of the Problems and of a Synthesis of the Results* (World Scientific, 2001).
35. Freire Jr, O. *et al. The Oxford Handbook of the History of Quantum Interpretations* (Oxford University Press, 2022).
36. Suzuki, M. General theory of fractal path integrals with applications to many-body theories and statistical physics. *Journal of Mathematical Physics* **32**, 400–407 (1991).
37. Suzuki, M. Fractal path integrals with applications to quantum many-body systems. *Physica A: Statistical Mechanics and its Applications* **191**, 501–515 (1992).
38. Kashiwa, T., Ohnuki, Y. & Suzuki, M. *Path Integral Methods* (Clarendon Press, 1997).
39. Koide, T. & Kodama, T. Generalization of uncertainty relation for quantum and stochastic systems. *Physics Letters A* **382**, 1472–1480 (2018).
40. Yasue, K. Stochastic calculus of variations. *Journal of Functional Analysis* **41**, 327–340 (1981).

41. Mandelstam, L. & Tamm, I. *The Uncertainty Relation Between Energy and Time in Non-relativistic Quantum Mechanics*, 115–123 (Springer Berlin Heidelberg, Berlin, Heidelberg, 1991).
42. Margolus, N. & Levitin, L. B. The maximum speed of dynamical evolution. *Physica D: Nonlinear Phenomena* **120**, 188–195 (1998).
43. Anandan, J. & Aharonov, Y. Geometry of quantum evolution. *Physical Review Letters* **65**, 1697 (1990).
44. Salecker, H. & Wigner, E. P. Quantum limitations of the measurement of space-time distances. *Part I: Particles and Fields. Part II: Foundations of Quantum Mechanics* 148–154 (1997).
45. Carruthers, P. & Nieto, M. M. Phase and angle variables in quantum mechanics. *Reviews of Modern Physics* **40**, 411 (1968).
46. Tanimura, S. Uncertainty relation between angle and orbital angular momentum: Interference experiment using electron vortex beams. *arXiv preprint arXiv:1411.0823* (2014).
47. Kochen, S. & Specker, E. P. The problem of hidden variables in quantum mechanics. *Journal of Mathematics and Mechanics* **17**, 59–87 (1967).
48. Einstein, A., Podolsky, B. & Rosen, N. Can quantum-mechanical description of physical reality be considered complete? *Physical Review* **47**, 777 (1935).

Summary

10

Abstract

This chapter gives the summary over all the contents in the book. Not only what have been clarified, but matters that remain mysterious are listed.

Before closing this book, I would like once again to quote the words allegedly stated by Feynman: "Quantum theory can be described, but it cannot be explained." However, no matter how extraordinarily gifted Feynman may have been, I believe that by steadily accumulating research efforts, even an ordinary researcher such as the present author can eventually expand the domain that can be understood and "explained." In concluding this book, I summarize what has become understandable and what remains as work yet to be pursued.

10.1 What Has Been Clarified: The Overall Structure and Interpretation of Schrödinger Dynamics

The principal aim of this book has not been to treat quantum mechanics as an algebraic structure, but rather to elucidate the dynamical structure of quantum systems based on the Schrödinger equation. This program consisted of three plans: (1) to derive the Schrödinger equation from scratch and clarify its meaning, as well as the role of the Schrödinger function and equation, (2) to formulate the quantum path integral at the level of real-valued functions in order to clarify the structure of the Schrödinger function as a sum over many events and to make explicit the integration measure, and (3) to regard each individual motion constituting the sum over events as a path in configuration space and to formulate its stochastic differential equation.

By explicitly formulating these three pillars, the essential features and nature of quantum theory at each level have been clarified. The following is a summary.

K. Takatsuka, *Structure of the Schrödinger Dynamics*,
https://doi.org/10.1007/978-981-92-1910-0_10

Derivation and Properties of the Real Schrödinger Equation

1. The distribution density function of quantum particles $\rho(q,t)$ is decomposed as $\rho(q,t) = \bar{\psi}(q,t)^T \bar{\psi}(q,t)$, using a two-component real-valued Schrödinger vector $\bar{\psi}(q,t)$ whose components are smooth multivariable space-time functions. In quantum theory, $\rho(q,t)$ and $\bar{\psi}(q,t)$ play roles at two hierarchical levels and are complementary in this sense.
2. $\bar{\psi}(q,t)$ is the distribution amplitude function of a particle ensemble.
3. By imposing the continuity equation on $\rho(q,t)$ using conservation laws and flux based on Gauss's theorem, and by requiring unitary space-time translation invariance, the formal structure of $\bar{\psi}(q,t)$ is established.
4. By taking the variation of the continuity equation for $\rho(q,t)$ with respect to $\bar{\psi}(q,t)$, the Schrödinger equation is obtained. The Schrödinger equation thus functions as a mathematical device that produces robust $\bar{\psi}(q,t)$ by making the continuity equation stationary.
5. $\bar{\psi}(q,t)$ is equivalent to the complex Schrödinger function. The presence of the imaginary unit corresponds to a rotation operation of the vector $\bar{\psi}(q,t)$.
6. The apparent wave-like nature of the Schrödinger equation arises from the linearity of the dynamics, and there is no need to introduce physically assumed matter waves, nor to invoke auxiliary equations such as the Hamilton-Jacobi equation.

Thus, the Schrödinger equation can be derived starting from the vector representation of the square root of $\rho(q,t)$ by following standard mathematical procedures, without intuitive leaps or inconsistencies. In this sense, Schrödinger theory is logically sound and natural. On the other hand, since quantum theory—including Schrödinger equation—has withstood experimental verification over the past century, it may also be regarded as correct as a "scientific theory," provided its domain of applicability is respected.

The distribution amplitude function $\bar{\psi}(q,t)$ in Schrödinger dynamics has two independent real values at each space-time point (q,t) and acts as the internal structure of the distribution function $\rho(q,t)$. There are infinitely many ways to assign these two values. Among them, in the polar representation $\left(\rho^{1/2}(q,t), \theta(q,t)\right)$, the spatial dynamics of the distribution are governed by $\vec{\nabla}\theta(q,t)$, the local energy is given by $\partial_t\theta(q,t)$, and the magnitude of the distribution is carried by $\rho^{1/2}(q,t)$.

Incidentally, since the Schrödinger function is a distribution amplitude function, it is reasonable to consider that there is no such thing as a "Schrödinger function of the universe" as appears in the many-worlds interpretation. (Here, the discussion is separated from cosmology and general relativity and is made purely as a matter of logical possibility.) This is because one would fall into a self-referential paradox in which the Schrödinger function that is supposed to describe the entire universe is itself a part of the distribution of that universe. (This is analogous to the paradox: "A Cretan says, 'All Cretans are liars.' Is this statement true or false?")

Setting-Up of Real-Valued Path Integral

1. A real-valued path integral was formulated as the Green function of the coupled column-vector-type real-valued Schrödinger equation for the Schrödinger vector.
2. Unlike the complex Feynman path integral, the real-valued path integral possesses a (skewed) Wiener measure.
3. The path integral clarifies the meaning of the Schrödinger function as a sum over paths, that is, a sum over events. Furthermore, it highlights the presence of stochasticity underlying quantum mechanics, and the close similarity with the Feynman-Kac formula in stochastic process theory and statistical mechanics strongly suggests that dynamics similar to that of Brownian particles should exist in quantum theory.

Formulation and Analysis of Quantum Stochastic Paths

1. An Ito-type stochastic differential equation for quantum stochastic paths was formulated.
2. The solution of this stochastic differential equation formally resembles that of classical Brownian motion; however, the velocity field coefficient is a functional of the Schrödinger function (in fact, the gradient of the angular component in the polar representation of the Schrödinger vector) and thus constitutes purely quantum dynamics.
3. The trajectories generated from the velocity field coefficient alone coincide with Bohm trajectories in the absence of the Wiener process and furthermore correspond to individual streamlines of the flux (manifold) generated by the Schrödinger vector. These naturally reproduce indirect correlations between quantum paths. The fringe patterns observed in the double-slit experiment is considered to originate from the non-crossing property of the underlying Bohm trajectories. The non-crossing rule is a reflection of the indirect correlation among the quantum paths.
4. The Wiener process associated with this stochastic differential equation yields various physical consequences. For example, uncertainty relations between position and momentum and between kinetic energy and time are obtained.
5. Quantum stochastic paths reflect the effects of quantum entanglement while also possessing the function of spontaneous disentanglement.
6. The Hamiltonian canonical equations and Newtonian equations of classical mechanics are reproduced as the limit $\hbar \to 0$ in the quantum canonical equations derived from the stochastic differential equation of quantum stochastic paths.

Combining these results, the following conclusions are obtained:

The Schrödinger function may be transformed into another functional form by experimental observation, but it does not physically collapse instantaneously. In the double-slit experiment, the traces of electrons left on the detection screen correspond to the landing points of individual quantum stochastic paths.

Moreover, for the distribution density function $\rho(q,t)$ to be regarded as a probability density function, an underlying stochastic process is required. One such element is the uncertainty relation. Therefore, it is impossible for us to specify experimental conditions in such a way that they can be reproduced exactly.

Furthermore, due to the Wiener process contained in the stochastic differential equation of quantum stochastic paths, the distribution $\rho(q,t)$ can only be reproduced probabilistically.

The Schrödinger function is a sum of infinitely many individual events. With the development of quantum technology, various experimental methods have advanced. However, in the analysis of experiments involving individual events or experiments that cannot be sufficiently repeated, one must employ quantum stochastic paths or theories based on them. Examples include few-collision experiments and single-molecule experiments at the attosecond scale.

Even when small changes in experimental conditions produce only apparently small fluctuations in the Schrödinger function as a whole, there can exist chaotic situations in which quantum stochastic paths undergo large variations. In such cases as well, analysis based on quantum stochastic paths is required. This demands further study as a new stage of quantum chaos.

In the formulation of quantum dynamics presented above, the "matter wave" as a physically real entity did appear nowhere in the present theory. The Schrödinger function is a distribution amplitude function, and the wave-like properties reflected in interference effects only arise from its linearity.

The imaginary unit contained in the Schrödinger function, which was initially thought to describe waves, is in fact associated with the rotation of the Schrödinger vector, as revealed by the derivation of the real-valued Schrödinger equation.

Even in quantum stochastic path dynamics, the "matter wave" does not appear as a physically observable entity. The velocity field coefficient is the gradient of the rotation angle of the Schrödinger vector and cannot itself be regarded as a physical wave.

10.2 Remaining Problems

It is believed that most of the issues raised in Chap. 3, particularly those summarized in Sect. 3.4, have been addressed. However, the following problems remain as issues that cannot be answered solely within the Schrödinger dynamics.

Dual Structure and Complementarity of $\rho(q,t)$ and $\bar{\psi}(q,t)$

In the relations $\rho(q,t) = \bar{\psi}(q,t)^T \bar{\psi}(q,t)$ and $\bar{\psi}(q,t) = \begin{pmatrix} \phi_r(q,t) \\ \phi_c(q,t) \end{pmatrix}$, the level we can directly recognize is $\rho(q,t)$, while the dynamics of the distribution function are determined by $\bar{\psi}(q,t)$ at the "square-root level." Particle-like properties are recognized at the level of $\rho(q,t)$, whereas the linearity of the distribution function (its wave-like aspect) appears mainly at the level of $\theta(q,t)$. Thus, the background of Bohr's concept of complementarity is also present here. However, why quantum theory—or our mode of cognition—has such a hierarchical structure remains unknown. At present, one can only say that "nature is so constituted." This is one of the ultimate problems that can be described but not explained.

Theory of Multicomponent Schrödinger Vectors

The nonrelativistic Schrödinger equation can be derived simply by imposing translational invariance in space-time on a two-component real vector. In other words, it is a theory that holds within a simple four-dimensional Euclidean space, albeit a coordinate transformation thereof. The fact that the Schrödinger equation involves second-order spatial derivatives and forms a closed equation is a consequence of imposing a conservation law for the flux in this space. In practice, when one attempts to incorporate Lorentz invariance or the curvature of space-time, it becomes readily apparent that an extension to vectors with multiple components is required.

Relation Between Parts and the Whole

As noted above, in the stochastic differential equation of quantum paths, the velocity field coefficient depends on the Schrödinger function. Therefore, it is understandable that each trajectory (except for the Wiener process) follows the distribution generated by the Schrödinger equation.

The velocity field coefficient can be regarded as a transmitting dynamical information determined at the level of $\bar{\psi}(q, t)$ (the whole) to dX_t at the higher level (the part) through Eq. (8.2.26). On the other hand, since the whole is constructed from the collection of the parts (via the path integral), the relation between parts and the whole becomes a self-referential nonlinear one.

Within quantum mechanics as developed so far, the Schrödinger equation and the Schrödinger function are solved first, once the initial and boundary conditions (including the potential function) are given, and quantum stochastic paths are then obtained via the Schrödinger function. In this sense, there is no mathematical inconsistency. (The same holds in diffusion theory.) However, from a physical standpoint, how can each individual path, which moves locally along the flux, sense the information of the Schrödinger function that extends over space-time?

The quantum canonical equations (9.1.19), written as a system of ordinary differential equations, provide an important clue for considering self-distribution dependence in quantum theory. According to this formulation, the presence of the trajectory density $\rho(X_t, t)$ multiplying the force term $\left(-\vec{\nabla} V(X_t, t)\right)$ makes it difficult to escape from the self-referential nature in which "one's own density determines the force acting on oneself."

It would be desirable if an element independent of the three pillars—analogous to the Langevin equation—could be found. However, the velocity field predicted by such a theory must correctly reflect dynamics arising from the principle of superposition, including entanglement, and therefore cannot be expressed merely in terms of local potential functions. Therefore, even if such a formulation exists in principle, its discovery would be extremely difficult.

Physical Origin of the Wiener Process
Another element that can be described mathematically but not explained physically is the physical origin of the Wiener process appearing in the stochastic differential equation of quantum paths.

From the correspondence between the Schrödinger equation and the diffusion equation, it is mathematically clear that stochastic motion is hidden in the background. It is also understood (mathematically) that the operations of taking infinitesimal in time and space are subject to constraints on their independence, which limit the smoothness of motion and give rise to uncertainty. Is there a physical background that brings about these constraints?

10.3 After All

If one pursues the mysteries of quantum mechanics to their ultimate extent, they are found to reduce, in the end, to the "remaining problems" summarized above: (1) that the summation of events is carried out not at the level of the distribution function $\rho(q, t)$ but at the level of the distribution amplitude function $\bar{\psi}(q, t)$, namely, the Schrödinger vector, (2) that each quantum trajectory evolves under the guidance of the global Schrödinger vector, and (3) that each quantum trajectory is accompanied by a Wiener process.

These are problems entrusted to readers who are interested in foundational issues, and it is hoped that they will pursue them further and in greater depth.

This book has not addressed applications at all. Nevertheless, readers engaged in cutting-edge research areas, such as quantum information, quantum measurement, quantum technology, and condensed matter physics, molecular science, and biological science, may have found hints toward applications scattered throughout the text. The author has enjoyed science for nearly fifty years within the realm of quantum mechanics, particularly within the world of the Schrödinger equation. It is the author's hope that this book may serve, even in a small way, as a return of that experience to younger researchers.

GPSR Compliance

The European Union's (EU) General Product Safety Regulation (GPSR) is a set of rules that requires consumer products to be safe and our obligations to ensure this.

If you have any concerns about our products, you can contact us on ProductSafety@springernature.com

In case Publisher is established outside the EU, the EU authorized representative is:

Springer Nature Customer Service Center GmbH
Europaplatz 3
69115 Heidelberg, Germany

Batch number: 10918189

Printed by Printforce, the Netherlands